Lecture Notes in Economics and Mathematical Systems

W0246148

Lecture Notes in Economics and Mathematical Systems

Kurt Marti (Ed.)

Stochastic Optimization

Numerical Methods and Technical Applications

Proceedings of a GAMM/IFIP-Workshop held at
the Federal Armed Forces University Munich,
Neubiberg, FRG, May 29–31, 1990

Springer-Verlag

Berlin Heidelberg New York
London Paris Tokyo
Hong Kong Barcelona
Budapest

Editor

Prof. Dr. Kurt Marti
Fakultät für Luft- und Raumfahrttechnik
Universität der Bundeswehr München
Werner-Heisenberg-Weg 39
W-8014 Neubiberg

ISBN-13: 978-3-540-55225-3   e-ISBN-13: 978-3-642-88267-8
DOI: 10.1007/978-3-642-88267-8

Typesetting: Camera ready by author

42/3140-543210 - Printed on acid-free paper

PREFACE

This volume includes a selection of papers presented at the GAMM/ IFIP-Workshop on "Stochastic Optimization: Numerical Methods and Technical Applications", held at the Federal Armed Forces University Munich, May 29-31, 1990.

The objective of this meeting was to bring together scientists from Stochastic Programming and from those Engineering areas, where Mathematical Programming models are common tools, as e.g. Optimal Structural Design, Power Dispatch, Acid Rain Abatement etc.. Hence, the aim was to discuss the effects of taking into account the inherent randomness of some data of these problems, i.e. considering Stochastic Programming instead of Mathematical Programming models in order to get solutions being more reliable, but not more expensive.

An international programme committee was formed which included

H.A. Eschenauer (Germany)

P. Kall (Switzerland)

K. Marti (Germany, Chairman)

J. Mayer (Hungary)

G.I. Schuëller (Austria)

Although the number of participants had to be small for technical reasons, the area covered by the lectures during the workshop was rather broad. It contains theoretical insight into stochastic programming problems, new computational approaches, analyses of known solution methods, and applications in such very different technical fields as ecology, energy demands, and optimal reliability of mechanical structures. In particular, the applied presentation also pointed to several open methodological problems.

In order to guarantee a high scientific level of the present Proceedings, all papers were refereed. Hence, we express our gratitude to all referees and to all contributors for delivering the final version of their papers in due time.

We gratefully acknowledge the financial support of GAMM (Gesellschaft für Angewandte Mathematik und Mechanik, IFIP (International Federation For Information Processing) and Federal Armed Forces University Munich.

Finally, we thank Springer-Verlag for including the Proceedings in the Springer Lectures Notes Series.

Munich

June 1991   K. Marti

CONTENTS

FINITE CONVERGENCE IN STOCHASTIC PROGRAMMING

Sjur D. Flåm[1]

Institute of Economics, Univ. of Bergen, 5008 Bergen, Norway.

ABSTRACT A differential inclusion is designed for solving stochastic, finite horizon, convex programs. Under a sharpness condition we demonstrate that the resulting method yields finite convergence.

Key words Convex program, saddle point, differential inclusion, stochastic programming, sharp constraints, finite convergence.

1. INTRODUCTION

This paper considers planning problems plagued by uncertainty about the (future) outcome ω in some event space Ω. Such problems can often be cast in the form of a constrained stochastic program

(P): Minimize the expected cost

$$(1.0) \quad f_0(x) := EF_0(\omega, x_1(\omega), .., x_S(\omega)) := \int F_0(\omega, x_1(\omega), .., x_S(\omega))\mu(d\omega)$$

with respect to the strategy profile $x = x(\cdot) = (x_1(\cdot), .., x_S(\cdot))$ under two types of constraints: First, we must cope with "technological" restrictions of the standard genre

$$(1.1) \quad F_s(\omega, x_1(\omega), .., x_S(\omega)) \leq 0 \quad \text{a.e. for} \quad s = 1, ..., S.$$

Here F_s takes values in $\mathbb{R}^{m_s}$, and (1.1), - which is understood to hold componentwise -, may reflect variable resource endowments, production possibilities and the like. Second, we face informational limitations expressed formally (and compactly) by

$$(1.2): \quad x_s(\cdot) \text{ should be } \Sigma_s\text{-measurable for } s = 1, ..., S.$$

---

[1] Written in parts at the Univ. of Bayreuth. The research has partially been supported by Ruhrgas via NAVF.

Two features are incorporated in (1.2): First, decisions are implemented <u>sequentially</u>. At each stage (decision epoch) $s = 1,2,..$ up to the planning horizon S included, an irreversible commitment $x_s(\omega) \in \mathbb{R}^{n_s}$ is made. Second, $x_s(\omega)$ is committed,- within a time window which opens temporarily at stage s -, under <u>imperfect</u> <u>information</u> about the exact state $\omega \in \Omega$ of the world. This stepwise resolution of uncertainty means, in more common jargon, that decisions never depend on <u>future</u> information. They are all <u>non-anticipative</u>, and resemble "sunk investments" once made: Historical decisions cannot be modified. By way of example, let the information flow be generated sequentially by a stochastic process $\xi_1,...,\xi_S$ on Ω. Then decision x_s cannot await either ξ_{s+1} or ξ_{s+2} ...or ξ_S. Rather, x_s should only take into account the actual realization of $\xi_1,...,\xi_S$. Thus, Σ_S is, in this case, the smallest σ-algebra rendering all (possibly vector) variates $\xi_1,...,\xi_S$ measurable.

It is also worthwhile to emphasize that all strategies $x_1(\cdot),..,x_S(\cdot)$ are laid down (computed) right <u>here</u> <u>and</u> <u>now</u>. This feature does not contradict the fact that one must <u>wait</u> <u>and</u> <u>see</u> (appropriate information) before these strategies can actually be implemented on line contingent upon how the system unfolds and uncertainty is unveiled.

This completes the heuristic description of the multi-stage stochastic optimization problem. Technical assumptions are relegated to Section 2. The purpose of this paper is to provide an algorithm, described in Section 3, which under broad hypotheses, yields finite convergence to optimal solutions. That algorithm amounts to simulate a very large scale, deterministic, differential system.

<u>2 PRELIMINARIES</u>   This section specifies the assumptions imposed on problem (P).

The operator E in (1.0) denotes the expectation over Ω, this set being conceived of as a probability space with sigma-algebra Σ and probability measure μ (possibly subjective).

We assume that Σ_s, $s = 1,2,...,S$, in (1.2) are complete sub-sigma-algebras of (Ω,Σ,μ).

Constraint (1.2) will be supplemented by requiring also

square integrability, i.e.,

(2.1) $x_s \in L^2(\Sigma, \mathbb{R}^{n_s})$ for all $s \geq 1$.

where $L^2(\Sigma, \mathbb{R}^{n_s})$ denotes the Hilbert space of square integrable, Σ-measurable random vectors in $\mathbb{R}^{n_s}$. In short, (1.2) and (2.1) say jointly that no strategy x can be selected outside the set

(2.2) $X := L^2(\Sigma_1, \mathbb{R}^{n_1}) \times \ldots\ldots \times L^2(\Sigma_S, \mathbb{R}^{n_S})$

We reiterate that (2.2) embodies two requirements: strategies must be non-anticipative and square integrable. Quite in line with (2.2), we demand that the common place "technological" restrictions (1.1) satisfy, for all $s \geq 1$, the two conditions

(2.3) $x \in X \implies F_s(\cdot, x_1(\cdot), .., x_s(\cdot))$ is Σ_s-measurable, and

(2.4) $x \in H \implies F_s(\cdot, x_1(\cdot), .., x_s(\cdot)) \in L^2(\Sigma, \mathbb{R}^{m_s})$ and continuous.

Here, for simplicity in notation, H denotes the Hilbert space $L^2(\Sigma, \mathbb{R}^n)$ with $n := n_1 + \ldots + n_S$.

Motivated by practical examples, and also by the need to remain within the confines of convex analysis, we posit that

(2.5)
the cost function $F_0(\omega, \cdot)$ and all m_s components of the constraint functions $F_s(\omega, \cdot)$, $s = 1, .., S$, are convex and finite-valued for all $\omega \in \Omega$.

Also, to make problem (P) tractable we have incorporated no constraints in the objective function f_0 (1.0). Specifically, we suppose that

(2.6) $f_0(x)$ is finite-valued continuous at all $x \in H = L^2(\Sigma, \mathbb{R}^n)$.

As customary, violations of (1.1) will be evaluated (or penalized) by means of multiplier vectors $y_s \in \mathbb{R}^{m_s}$, $s = 1, \ldots, S$. These multipliers are random however [2]. Specifically, in accord with (2.3) and (2.4), we posit that all $y_s(\cdot)$ be Σ_s-measurable, square integrable. For notational convenience, we shall codify this requirement by saying that any multiplier $y = (y_1,\ldots,y_S)$ must belong to the Hilbert space

(2.7) $Y := L^2(\Sigma_1,\mathbb{R}^{m_1}) \times \cdots \times L^2(\Sigma_S,\mathbb{R}^{m_S})$.

Such multipliers $y \in Y$ enter into a "functional" <u>Lagrangian</u>

(2.8) $L(x,y) := \int \Lambda(\omega,x(\omega),y(\omega))d\mu(\omega)$.

where the integrand $\Lambda:\Omega\times\mathbb{R}^n\times\mathbb{R}^m \to \mathbb{R}$ is a "pointwise" Lagrangian

(2.9) $\Lambda(\omega,\xi,\eta) := F_0(\omega,\xi) + \displaystyle\sum_{s=1}^{S} \eta_s \cdot f_s(\omega,\xi_1,\ldots,\xi_S)$

defined for all $\xi = (\xi_1,\ldots,\xi_S) \in \mathbb{R}^n$, $n = n_1 + \ldots + n_S$, and all $\eta = (\eta_1,\ldots,\eta_S) \in \mathbb{R}^m$, $m := m_1 + \ldots + m_S$. A non-standard feature appears in (2.9): The function $f := (f_s)_{s=1}^{S} := F^+ := (F_s^+)_{s=1}^{S}$ mentioned there is a shorthand for the positive part

(2.10) $f_s(\omega,\cdot) := \text{Max } \{0,F_s(\omega,\cdot)\}$ a.e.

the maximum operation in (2.10) being taken both pointwise and coordinatewise. More generally, in (2.9) we can let

$f_s(\omega,\cdot) := \varphi_s(\text{Max } \{0,F_s(\omega,\cdot)\})$ a.e.

with $\varphi_s :\mathbb{R}_+^{m_s} \to \mathbb{R}_+^{m_s}$ non-decreasing convex and vanishing only at the origin. The only essential restriction here is that we want the implication

(2.11) $x \in X \implies f_s(\cdot,x_1(\cdot),\ldots,x_S(\cdot)) \in L^2(\Sigma_S,\mathbb{R}^{m_S})$

to hold for all $s \geq 1$, as indeed it does under (2.3-4) and (2.10). To reiterate: The non-conventional property of the Lagrangian L in (2.8-10) is that only strict constraint violations are priced by means of multipliers. No gain is obtained by slackness. In other words, what we invoke is a (one-sided) exterior penalty method employing non-standard multipliers. Moreover, according to (2.7) these multipliers must be non-anticipative and square integrable. As customary, only non-negative multipliers are of interest, i.e., we shall invariably select them from the cone

(2.12) $Y_+ := \{y \in Y : y \geq 0 \text{ a.e.}\}.$

        Observe, via (2.3-4) and (2.6-10)), that the integral in (2.8) defines a finite, bivariate, function L over the space $H \times Y$. Furthermore, by the convexity assumption (2.5), this function L is convex-concave on $H \times Y_+$. Not surprisingly, L will be our main object in searching for solutions to problem (P).

3. THE PRIMAL-DUAL DIFFERENTIAL METHOD   We are now prepared to state our algorithm. To solve problem (P) we propose to follow a trajectory $(x,y)(t,\omega)$, $t \geq 0$, $\omega \in \Omega$, of the differential inclusion

(DI)
$$\dot{x}(t) \in -\Pi_X \partial_x L(x(t),y(t))$$
$$\dot{y}(t) \in \partial_y L(x(t),y(t))$$

verifying the viability condition: $y(t) \geq 0$ a.e. for all $t \in \mathbb{R}_+$.

Here $\dot{x}(t),\dot{y}(t)$ denote the time derivative, L was defined in (2.8-10), and by a trajectory we mean an absolutely continuous function $(x,y)(\cdot) : \mathbb{R}_+ \to H \times Y$ satisfying (DI) almost everywhere (a.e.). In the first inclusion here above Π_X signifies the orthogonal projection onto the set X (2.2). Also, in (DI), the partial subgradient operators ∂_x, ∂_y should be understood in the sense of convex analysis [7]. To wit,

$$\partial_x L(x,y) := \partial[L(\cdot,y)](x) = \partial f_0(x) + \partial\langle y,f(\cdot)\rangle(x),$$

$$\partial_y L(x,y) := -\partial[-L(x,\cdot)](y) = \{f(x)\}.$$

The dynamics (DI) can be interpreted as a continuous (infinite-simal) steepest feasible direction method in both variables x and y separately. It also portrays a process driven by first-order, myopic adjustments made by two zero-sum players [3]. Observe that the projection operator Π_X in (DI) takes care of one viability concern, namely that $x(t) \in X$, for all $t \in \mathbb{R}_+$. The other concern, that $y(t) \geq 0$ a.e., has no bite here. It is automatically satisfied as long as the initial guess y(0) is nonnegative a.e.

To make (DI) handy in computations we must evaluate the partial subdifferentials $\partial_x L(x,y)$ and $\partial_y L(x,y)$. When $y \in Y_+$ as defined in (2.12), a general rule for computing subgradients of convex integral functionals (cf. [6] and [9,p.442]), yields

(3.1) $\partial_x L(x,y) = \{u \in H : u(\omega) \in \partial_\xi \Lambda(\omega, x(\omega), y(\omega))$ a.s.$\}.$

In (3.1) the partial subdifferential $\partial_\xi \Lambda$ of Λ with respect to $\xi = (\xi_1, \ldots, \xi_S)$ can be evaluated directly from (2.9-10). Similarly, one gets

(3.2) $\partial_y L(x,y) = \{y \in Y : y(\omega) \in \partial_\eta \Lambda(\omega, x(\omega), y(\omega))$ a.e.$\}$
$= \{y \in Y : y_s(\omega) = f_s(\omega, x_1(\omega), \ldots, x_s(\omega)$ a.e.
for all $s = 1, \ldots, S\}.$

We shall see shortly that the representations of (3.1-2) of the partial subdifferentials take us a long way to make (DI) tractable. First, we need however, to spell out the projection operator Π_X in (DI). For this, recall that all sigma-algebras Σ_s (s=1,...,S) are complete. Then, evidently, X is a closed linear subspace of H. But, orthogonal projection of H onto $H_s := L^2(\Sigma_s, \mathbb{R}^{n_s})$ amounts to conditional expectation with respect to Σ_s, i.e.,

$$\pi_X[x] = (\pi_{H_1}[x], \ldots \ldots, \pi_{H_S}[x])$$

with

$$\pi_{H_S}[x] = E_S[x_S] := E(x_S|\Sigma_S) \quad \text{for all } x = (x_1,\ldots,x_S) \in H.$$

Thus, by this last observation and formulas (3.1-2), the "functional" differential inclusion (DI) splits finally into a system of "pointwise" inclusions

$(DI)_\omega$
$$\dot{x}_S(t)(\omega) \in -E_S[\partial_{\xi_S}\Lambda(\omega,x(t)(\omega),y(t)(\omega))]$$
$$\dot{y}_S(t)(\omega) = f_S(\omega,x_1(t)(\omega),\ldots,x_S(t)(\omega)).$$

In computations this latter system $(DI)_\omega$ is the one which should be solved (integrated numerically) for all stages $s = 1,\ldots,S$ and for almost every $\omega \in \Omega$. Any $(x(0),y(0))$ in the set $X \times Y_+$ (see (2.2) and (2.12)) can be used as initial point. Clearly, system $(DI)_\omega$, $\omega \in \Omega$, may be very large. Therefore, in practice one may have to contend with discrete probability measures approximating the original μ. Computations are however, beyond the scope of this paper. Rather, in the next section we only explore the convergence properties of (DI).

4 CONVERGENCE This section contains all novelties of this paper. It is concerned with the theoretical efficiency of (DI) as a computational method. Specifically, we shall show, under broad hypothesis, that (DI) can be expected to yield convergence in finite time. Our analysis adds to the following result of Flåm and Seeger [5].

THEOREM 4.1 (Convergence) Suppose that the set S of saddle points of L with respect to $X \times Y_+$ is nonempty. Then, any trajectory $(x,y)(\cdot)$ of (DI) emanating from $X \times Y_+$ is bounded and converges monotonically in norm to S. This trajectory stays within $X \times Y_+$, and $y(t)$ converges weakly and monotonically upwards to some $\bar{y} \in Y_+$. ∎

Brief outline of proof Let $L(x,y) = -\infty$ whenever $y \notin Y_+$, and $+\infty$ whenever $x \notin X$, $y \in Y_+$. Write $z = (x,y)$. Then the correspondence $M(z) := (\partial_x L, -\partial_y L)(z)$ is maximal monotone [8].

Consequently the "gradient" system

$$\dot{z}(t) \; \epsilon - M(z(t)), \quad z(0) \; \epsilon \; X \times Y_+ \quad \text{specified,}$$

admits a unique, infinitely extendable, bounded trajectory which
lives in $X \times Y_+$ forever [1]. This trajectory also solves
(DI). Since $y(\cdot)$ is bounded and monotone, it converges weakly
and monotonically upwards to some $\bar{y} \; \epsilon \; Y_+$. Next, all weak
accumulations points of $x(t)$, $t \; \epsilon \; \mathbb{R}_+$ must be feasible, because
otherwise $\|y(t)\| \uparrow +\infty$. Finally, consider the Lyapunov function

$$\lambda(t) \; := \; \text{dist}(x(t),S_P)^2/2 \; + \; \|y(t) - \bar{y}\|^2/2$$

where $\text{dist}(\cdot,S_P)$ denotes the distance to the set S_P of
optimal solutions to problem (P). The right hand derivative of
$\lambda(t)$ is majorized by $\inf(P) - f_o(x(t))$. Therefore, $\lambda(t) \downarrow -\infty$
unless all weak accumulation points of $x(t)$, $t \; \epsilon \; \mathbb{R}_+$, belongs to
S_P. ∎

We desire here not only convergence as ensured by Thm.4.1, but
also that this occur in finite time. For that purpose we need an

ASSUMPTION (On sharpness of constraints)
We say that the constraints of problem (P) are sharp with
modulus $\alpha > 0$ if

(4.1) $\mathbf{E1}\cdot f(x) \; \geq \; \alpha\,\text{dist}(x,C)$ for all $x \; \epsilon \; X$.

Here X is as defined in (2.2); C denotes the feasible set for
problem (P);
$$\text{dist}(x,C) \; := \; \inf_{c \epsilon C}\|x - c\|$$

is the L^2-distance between $x \; \epsilon \; X$ and C; the constant vector
$\mathbf{1} = (1,\ldots,1)$ belongs to $\mathbb{R}^m$, $m = m_1 + \ldots + m_S$; and finally,
$f = (f_s)_{s=1}^{S}$.

THEOREM 4.2 (Feasibility in finite time)

Suppose all coordinates of the initial multipliers $y_s(0)$, s = 1,.., S are minorized a. e. by a positive number γ. Also suppose constraints are sharp with modulus $\alpha > 0$, and that

$$\sup_{x \in B} \sup_{g_o \in \partial f_o(x)} \|g_o\| - \alpha\gamma \; =: \; -\Delta \; < \; 0,$$

where B is a ball centered at the optimal solution $\bar{x}(0)$ nearest to x(0) with radius majorizing $\|(x,y)(0) - (\bar{x},\bar{y})(0)\|$. Then x(t) is feasible for all   t $\geq$ dist (x(0),C)/Δ.

PROOF In Flåm & Seeger [5] it is shown that

$$\|x(t)-\bar{x}(0)\| \; \leq \; \|(x,y)(t)-(\bar{x},\bar{y})(0)\| \; \leq \; \|(x,y)(0)-(\bar{x},\bar{y})(0)\|.$$

Therefore x($\cdot$) stays within the ball B mentioned here above. Consider the distance $\delta(t) := \text{dist}(x(t),C)$ between the current point x(t) and the feasible set C.   As long as x(t) $\notin$ C we have

$$\delta(t)\dot{\delta}(t) = d(\delta(t)^2/2)/dt$$

$$= \langle x(t)-x^*(t), \dot{x}(t) \rangle$$

(where the derivative is taken from the right, and where $x^*(t)$ denotes the feasible point which is closest to x(t))

$$= E \sum_{s=0}^{S} \langle x(t)-x^*(t),-g_s(t) \rangle$$

(for $y_o \equiv 1$ and appropriate subgradients

$$g_s(t) \in \partial_x[y_s(t) \cdot f_s(x(t))], \; s = 0,\ldots,S)$$

$$\leq \delta(t)\|g_o(t)\| - E\sum_{s=1}^{S} y_s(t) \cdot f_s(x(t))$$

$$\leq \delta(t)\|g_o(t)\| - \gamma \, E\mathbb{1} \cdot f(x(t))$$

$$\leq \delta(t)[\|g_o(t)\| - \alpha\gamma] \; \leq \; -\delta(t)\Delta.$$

It follows that

$$\dot{\delta}(t) \leq - \Delta, \quad \text{when} \quad x(t) \notin C,$$

and now the conclusion is immediate. ∎

REMARK Thus, to obtain feasibility in a finite lapse of time one should choose all initial values $y_1(0), \ldots, y_S(0)$ large a.e. Conceptually one might contemplate to set $y_s(0) = +\infty$ for all $s \geq 1$. In practice, this is impossible however, and large values $y_s(0)$ may yield a fairly stiff system. ∎

In the light of Thm. 4.2 it is natural to inquire when constraints are indeed sharp. The next result, inspired by [4], gives a sufficient condition in this direction. For its statement some notation is needed. We introduce the cone

$$Y_{S+} := \{y_S \in L^2(\Sigma_S, \mathbb{R}^{m_S}): y_S \geq 0 \text{ a.e.}\}$$

of non-negative, Σ_S-measurable, square integrable random vectors in $\mathbb{R}^{m_S}$. Let the correspondence $G = (G_S)_{s=1}^S$ from X (2.2) to Y (2.7) be defined by

$$G_S(x) := F_S(\cdot, x_1(\cdot), \ldots, x_S(\cdot)) + Y_{S+}.$$

Note that feasibility of x in problem (P) amounts to the statement that $0 \in G(x)$. Thus, the feasible set C equals $G^{-1}(0)$. Recall that any L^2 space of (square integrable) random vectors may be regarded as a subset of the corresponding space L^1 of absolutely summable random vectors. Thus, on L^2 we also have a relative topology induced by the L^1-norm.

PROPOSITION 4.1 (Sharp constraints)
<u>Suppose the range of the correspondence</u> G <u>contains the origin as interior point and is closed in the</u> L^1<u>-completion of</u> $L^2(\Sigma, \mathbb{R}^m)$. <u>Then the constraints of problem</u> (P) <u>are sharp on any bounded set.</u>

PROOF   On the L^1-completion of the range space of   G, which is
Banach,   we temporarily use the L^1-norm, and denote it by $\|\cdot\|_1$.
Observe, using this norm, that

$$\text{dist}(G(x),0) = \|F^+(x)\|_1 = \|f(x)\|_1 = E\mathbb{1}\cdot f(x)$$

for every x ε X.   Now, by the Robinson-Ursescù theorem [1],   for
any x_o ε $G^{-1}(0)$ there exists   γ > 0   such that

$$\text{dist}(x,C) = \text{dist}(x,G^{-1}(0)) \leq \text{dist}(G(x),0)(1 + \|x-x_o\|)/\gamma$$

$$= E\mathbb{1}\cdot f(x)(1 + \|x-x_o\|)/\gamma \leq E\mathbb{1}\cdot f(x)(1 + \|x\| + \|x_o\|)/\gamma$$

for every   x ε X.   The conclusion is then immediate provided
all vectors   x   in question are uniformly bounded in norm.   ∎

REMARKS   (i) Suppose   Σ   is finite (so that   μ   has finite
support). Then constraints are sharp under the Slater condition
requiring that   (P) be strictly feasible, i.e., there should
exist   x ε X   such that (1.1) holds with strict inequality in
every coordinate.   In this case the hypothesis of Prop.4.1 is
satisfied.

   (ii) The conditions imposed in Prop. 4.1 are very
strong. Essentially, they imply that   Σ   is finite, so that the
L^1- and the L^2-topologies coincide.   Otherwise, when   Σ   contains
a sequence of events A_k, k = 1,2,..   such that $\mu(A_k)$ is
strictly decrasing to zero, one may easily show that L^2 is not
closed in L^1.

   (iii) The most important practical instances of   (P)
are linearly constrained.   Then (1.1) reads

$$A_s(\omega)[x_1(\omega),\ldots,x_S(\omega)]^T \leq b(\omega) \text{ a.e. for } s = 1,\ldots,S.$$

The possibly random technology matrix $A(\omega) = [A_1(\omega),\ldots,A_S(\omega)]$
defines here a linear mapping

$$(Ax)_s(\omega) := A_s(\omega)[x_1(\omega),\ldots,x_S(\omega)]^T$$

from   X (2.2)   into   Y (2.7).   Using the so-called Hoffmann inequality one may then show, again provided Σ is finite, that constraints are sharp. ∎

Once we have obtained feasibility, or even before, it is time to worry about optimality. To this end consider the derivative

$$f_0'(x;d) := \lim_{h \downarrow 0} \frac{f_0(x+hd)-f_0(x)}{h}$$

in any direction

(4.2) $\qquad d \in D := \Pi_X[\ - \sum_{s=0}^{S} \partial_X[y_s \cdot f_s(x)]]\ ,$

as prescribed by (DI). To reduce f_0 swiftly it is safe to select a direction

$$\dot{x} \in \text{argmin}_{d \in D}\ f_0(x;d).$$

Such a choice yields a <u>directional</u> <u>derivative</u>

$$f_0'(x;\dot{x}) = \min_{d \in D}\ f_0'(x;d)$$

$$= \min_{d \in D} \quad \max_{g_0 \in \partial f_0(x)} \quad \langle g_0, d \rangle$$

In particular, when $x(t)$ is feasible, we may select the direction $d(t)$ such that the contribution from every term $\partial_X[y_s(t) \cdot f_s(x(t))]$, $s \geq 1$, in (4.2) is nil. It follows then that

$$f_0(x,\dot{x}) \leq -\|g_0\|^2 \quad \text{for all} \quad g_0 \in \partial f_0(x).$$

To reflect this we say that f_0 <u>descends</u> <u>at</u> <u>least</u> <u>linearly</u> <u>on</u> C if $x \in C$ implies

$$f_0(x,\dot{x}) \leq -\|g_0\|^2 \quad \underline{\text{for}}\ \underline{\text{some}} \quad g_0 \in \partial f_0(x).$$

<u>THEOREM 4.3</u> (Finite convergence)

Suppose x(·) generated by (DI) is feasible for all t ≥ some
τ ≥ 0. Also, suppose that problem (P) is essentially constrained in the sense that

$$\mu := \inf_{x \epsilon C, g_o \epsilon \partial f_o(x)} \|g_o\|^2 > 0.$$

Then, if f_o descends at least linearly on C, x(t) is optimal
no later than time

$$t = [f_o(x(\tau)) - \inf(f_o | C)]/\mu + \tau.$$

PROOF When x(t) ε C we have

$$\frac{df_o(x(t))}{dt} = f_o'(x(t); \dot{x}(t))$$
$$\leq -\|g_o(t)\|^2 \quad (\text{for some } g_o(t) \epsilon \partial f_o(x(t))) \leq -\mu.$$

Hence, before optimality has occured it holds that

$$f_o(x(t)) - f_o(x(\tau)) \leq -\mu(t - \tau)$$

for all t ≥ τ, whence the conclusion is immediate. ∎

5 CONCLUDING REMARKS Stochastic programming is quite
challenging: Neither modeling nor computation is straight-
forward. As regards the latter issue most effort has naturally
been directed towards decomposition in one form or another [10].
Here we have gone very far in that direction: Problem (P) is
explored by means of a very large scale differential system.
That system updates all decisions (primal variables) and multipliers simultaneously. If data are smooth, the system dynamics
(DI) involve "kinks" which are "few" and easy to handle.
Moreover, it is only the asymptotic behavior of (DI) which is of
interest. It is comforting then that (DI) enjoys good stability
properties provided constraints are sharp.

REFERENCES

[1] J.P. Aubin and A. Cellina. Differential Inclusions,

Springer Verlag, N.Y. (1984).

[2] S.D. Flåm. "Lagrange multipliers in stochastic programmming", Preprint, Inst. of Economics, Univ. of Bergen (1990).

[3] S.D. Flåm and A. Ben-Israel. "Approximating Saddle Points as Equilibria of Differential Inclusions", JOTA 141, (1989), 264-277.

[4] S.D. Flåm and J. Zowe. "A Primal-Dual Differential Method for Convex Programming". Preprint, Institute of Economics, Univ. of Bergen, (1990).

[5] S.D.Flåm and Alberto Seeger. "Solving cone-constrained convex programs by differential inclusions", Working Paper, Institute of Economics, University of Bergen (1990).

[6] R.T. Rockafellar. "Integrals wich are Convex Functionals", Pacific J. of Math 24, (1968), 525-539.

[7] R.T. Rockafellar. Convex Analysis. Princeton Univ. Press, Princeton, N.J., (1970).

[8] R.T. Rockafellar. "Monotone Operators Associated with Saddle-Functions and Minimax Problems". In Nonlinear Functional Analysis (Proc. of Symposia in Pure Mathematics, Vol. 18, Part 1), F.E. Browder, Ed., (1970), 241-250.

[9] R.T. Rockafellar. "Integrals which are convex functionals", Part II. Pacific J. of Math 39, (1971), 439-469.

[10] R.T.Rockafellar and R.J-B.Wets. "Generalized linear-qua dratic problems of deterministic and stochastic control in discrete time". SIAM J. Control and Optimization, vol. 28, no. 4, 810-822 (1990).

LATTICE RULES FOR MULTIPLE INTEGRATION

HARALD NIEDERREITER

Institute for Information Processing
Austrian Academy of Sciences
Sonnenfelsgasse 19
A-1010 Vienna, Austria
E-mail: nied@qiinfo.oeaw.ac.at

1. Introduction

Numerical methods for stochastic optimization often involve as basic steps the calculation of probabilities and of expected values in a multivariate setting. For this purpose, efficient routines for multidimensional numerical integration are needed. We discuss here a class of quasi-Monte Carlo methods which have been developed recently and which yield powerful tools for multiple integration. For a general background on quasi-Monte Carlo methods we refer to the survey articles of Niederreiter [12], [15] and to the forthcoming book [18].

In the s-dimensional case we normalize the integration domain to be the s-dimensional unit cube $U^s = [0,1]^s$ (in Section 4 we will also consider compact groups as integration domains). For $s = 1$ we have *classical integration rules* such as the trapezoidal rule and Simpson's rule. For instance, the trapezoidal rule uses the nodes $n/m, n = 0, 1, \ldots, m$, with well-known weights and leads to an integration error $O(m^{-2})$ provided the integrand has a continuous second derivative. The classical rules can be extended to the multidimensional case $s \geq 2$ by a cartesian product construction. For instance, for the multidimensional trapezoidal rule we get the nodes

$$\left(\frac{n_1}{m}, \ldots, \frac{n_s}{m}\right)$$

with the n_j running independently through the set $\{0, 1, \ldots, m\}$. The total number of nodes is $N = (m+1)^s$. Under a suitable regularity assumption, the integration error is still $O(m^{-2})$; the fact that no improvement on the one-dimensional error bound is possible can be seen by considering an s-dimensional integrand which depends only on one variable. In terms of the number N of nodes, the error bound is thus of the order of magnitude $N^{-2/s}$. Therefore, this method becomes useless for high-dimensional integrals. Similar statements hold for cartesian products of other one-dimensional integration rules.

A method which has been used extensively for high-dimensional integrals is the *Monte Carlo method*. In this well-known stochastic method one takes N independent random samples $\mathbf{x}_1, \ldots, \mathbf{x}_N$ from the uniform distribution on U^s and uses the approximation

$$\int_{U^s} f(\mathbf{t})d\mathbf{t} \approx \frac{1}{N} \sum_{n=1}^{N} f(\mathbf{x}_n). \tag{1}$$

For square-integrable f the expected integration error is $O(N^{-1/2})$, an order of magnitude independent of the dimension s. On the other hand, this error bound is probabilistic, and so there is no guarantee that it can be achieved with a concrete choice of sample points. Another difficulty with the Monte Carlo method is the actual generation of independent random samples.

These difficulties can be overcome by applying *quasi-Monte Carlo methods*, which may be described as deterministic analogs of the Monte Carlo method. We again use the approximation (1), but with deterministic nodes $\mathbf{x}_1, \ldots, \mathbf{x}_N \in U^s$. Then we have a deterministic error bound given by the Koksma–Hlawka inequality (see [8, Ch. 2], [12, Sect. 2]), which says that if f has bounded variation $V(f)$ on U^s in the sense of Hardy and Krause, then

$$|\frac{1}{N}\sum_{n=1}^{N} f(\mathbf{x}_n) - \int_{U^s} f(\mathbf{t})dt| \leq V(f)D_N, \qquad (2)$$

where the *discrepancy* $D_N = D_N(\mathbf{x}_1, \ldots, \mathbf{x}_N)$ measures the irregularity of distribution of the nodes. In detail, we have

$$D_N = \sup_J |\frac{1}{N}\#\{1 \leq n \leq N : \mathbf{x}_n \in J\} - \mathrm{Vol}(J)|,$$

where the supremum is extended over all intervals $J = \prod_{j=1}^{s}[0, u_j] \subseteq U^s$. For efficient quasi-Monte Carlo methods we need nodes $\mathbf{x}_1, \ldots, \mathbf{x}_N$ with small discrepancy D_N, and such sets of nodes are called *low-discrepancy point sets*. The aim is to get low-discrepancy point sets with $D_N = O(N^{-1}(\log N)^{e(s)})$, where the exponent $e(s)$ depends only on s. The best constructions of low-discrepancy point sets (see Sobol' [26], Faure [2], and Niederreiter [14], [16]) achieve $e(s) = s - 1$, and this gives integration rules that are more efficient than the Monte Carlo method. However, the calculation of the nodes in these constructions is rather time-consuming.

This raises the important question whether there are low-discrepancy point sets which are easy to generate. The answer is affirmative provided that we are willing to make $e(s)$ slightly larger, say $e(s) = s$ will do. Suitable low-discrepancy point sets are obtained as the node sets of so-called lattice rules which we are going to discuss. It will also transpire that lattice rules yield much smaller error bounds in the case of periodic integrands.

2. Basic theory of lattice rules

An s-dimensional *lattice* is obtained by taking s linearly independent vectors $\mathbf{y}_1, \ldots, \mathbf{y}_s \in \mathbf{R}^s$ and forming the set $L = \{\sum_{j=1}^{s} a_j \mathbf{y}_j : a_j \in \mathbf{Z}\}$ of all integer linear combinations. We only consider lattices L containing the integer lattice $\mathbf{Z}^s$. Then $L \cap [0, 1)^s$ is a finite set, say $\{\mathbf{x}_1, \ldots, \mathbf{x}_N\}$. This finite set is called the *node set* of the *lattice rule* L, where the latter uses these nodes in the approximation (1). If we want to emphasize that the number of nodes in a lattice rule is N, then we speak of an *N-point lattice rule*. To exclude a trivial case, we always assume $N \geq 2$. The first steps in the direction of lattice rules were taken by Frolov [3] and a systematic theory was developed by Sloan [21] and Sloan and Kachoyan [22], [23]. Special classes of lattice rules were introduced much earlier by Korobov [7] and Hlawka [5].

For periodic integrands the integration error in a lattice rule L can be expressed as follows. Let f be periodic with period interval U^s and suppose that f is represented by the absolutely convergent Fourier series

$$f(\mathbf{t}) = \sum_{\mathbf{h} \in \mathbf{Z}^s} \widehat{f}(\mathbf{h})e^{2\pi i \mathbf{h} \cdot \mathbf{t}}$$

with Fourier coefficients $\widehat{f}(\mathbf{h})$, where $\mathbf{h} \cdot \mathbf{t}$ denotes the standard inner product of $\mathbf{h}$ and $\mathbf{t}$. Define the *dual lattice*

$$L^\perp = \{\mathbf{h} \in \mathbf{Z}^s : \mathbf{h} \cdot \mathbf{x} \in \mathbf{Z} \quad \text{for all} \quad \mathbf{x} \in L\}.$$

Then by [23, Theorem 2] we have

$$\frac{1}{N}\sum_{n=1}^{N} f(\mathbf{x}_n) - \int_{U^s} f(\mathbf{t})dt = \sum_{\substack{\mathbf{h}\in L^{\perp} \\ \mathbf{h}\neq 0}} \widehat{f}(\mathbf{h}). \tag{3}$$

We introduce the following regularity classes for periodic functions.

Definition 1. Let $\alpha > 1$ and $A > 0$ be real. Then $f \in \mathcal{E}_\alpha^s(A)$ if f is continuous, has U^s as its period interval, and its Fourier coefficients satisfy

$$|\widehat{f}(\mathbf{h})| \leq Ar(\mathbf{h})^{-\alpha} \quad \text{for all nonzero} \quad \mathbf{h} \in \mathbf{Z}^s,$$

where for $\mathbf{h} = (h_1, \ldots, h_s) \in \mathbf{Z}^s$ we put

$$r(\mathbf{h}) = \prod_{j=1}^{s} \max(1, |h_j|).$$

We write $f \in \mathcal{E}_\alpha^s$ if $f \in \mathcal{E}_\alpha^s(A)$ for some $A > 0$.

A sufficient condition for membership in such a regularity class which is easier to check is the following: if $\alpha \geq 2$ is an integer, the partial derivative $\partial^{\alpha s} f/\partial t_1^\alpha \ldots \partial t_s^\alpha$ is continuous, and f has U^s as its period interval, then $f \in \mathcal{E}_\alpha^s$. For real $\alpha > 1$ put

$$R_\alpha(L) = \sum_{\substack{\mathbf{h}\in L^{\perp} \\ \mathbf{h}\neq 0}} r(\mathbf{h})^{-\alpha}.$$

Then it follows from (3) and Definition 1 that if $f \in \mathcal{E}_\alpha^s(A)$, then

$$\left|\frac{1}{N}\sum_{n=1}^{N} f(\mathbf{x}_n) - \int_{U^s} f(\mathbf{t})dt\right| \leq AR_\alpha(L). \tag{4}$$

$R_\alpha(L)$ can be bounded in terms of the *figure of merit*

$$\varrho(L) = \min_{\substack{\mathbf{h}\in L^{\perp} \\ \mathbf{h}\neq 0}} r(\mathbf{h}).$$

In fact, we have

$$\varrho(L)^{-\alpha} \leq R_\alpha(L) = O\left(\varrho(L)^{-\alpha}(1 + \log \varrho(L))^{s-1}\right),$$

where the lower bound is trivial and the upper bound was shown by Sloan and Kachoyan [23]. Thus $\varrho(L)$ should be as large as possible to get an efficient integration rule.

An error bound for nonperiodic integrands in a lattice rule is obtained from (2), but we need bounds for the discrepancy of the node set of a lattice rule L. We write $D(L)$ for the discrepancy $D_N(\mathbf{x}_1, \ldots, \mathbf{x}_N)$ of the node set. We put $C_s(N) = ((-N/2, N/2] \cap \mathbf{Z})^s, C_s^*(N) = C_s(N) \setminus \{0\}, E(L) = C_s^*(N) \cap L^{\perp}$, and

$$R_1(L) = \sum_{\mathbf{h}\in E(L)} r(\mathbf{h})^{-1}.$$

Then according to a result of Niederreiter and Sloan [19] we have

$$D(L) \leq \frac{s}{N} + \frac{1}{2} R_1(L). \tag{5}$$

In [19] it was also shown that the discrepancy $D(L)$ can be bounded in terms of the figure of merit by means of the inequalities

$$c_s \varrho(L)^{-1} \leq D(L) \leq c'_s \varrho(L)^{-1} (\log N)^s,$$

where the positive constants c_s, c'_s depend only on s.

The theory of lattice rules also contains an interesting group-theoretic aspect. If $\{\mathbf{x}_1, \ldots, \mathbf{x}_N\}$ is the node set of a lattice rule L, then the residue classes $\mathbf{x}_1 + \mathbf{Z}^s, \ldots, \mathbf{x}_N + \mathbf{Z}^s$ form the finite subgroup $L/\mathbf{Z}^s$ of the torus group $\mathbf{R}^s/\mathbf{Z}^s$. By using the structure theory for finite abelian groups, Sloan and Lyness [24] have shown the following fundamental classification theorem.

Theorem A. *For any N-point lattice rule in dimension s there exist a uniquely determined integer r with $1 \leq r \leq s$ and uniquely determined integers $n_1, \ldots, n_r > 1$ with n_{i+1} dividing n_i for $1 \leq i < r$ such that the node set consists exactly of all fractional parts*

$$\left\{ \sum_{i=1}^{r} \frac{k_i}{n_i} \mathbf{z}_i \right\} \quad with \quad 1 \leq k_i \leq n_i \quad for \quad 1 \leq i \leq r,$$

where $\mathbf{z}_1, \ldots, \mathbf{z}_r \in \mathbf{Z}^s$ are linearly independent. Also $N = n_1 \ldots n_r$.

Definition 2. The integer r in Theorem A is called the *rank* and the integers $n_1, \ldots, n_r$ in Theorem A are called the *invariants* of the lattice rule.

Theorem A yields general bounds for the various efficiency measures introduced above. These bounds involve the first invariant n_1.

Theorem 1. *For any s-dimensional N-point lattice rule L we have:*

(i) $\varrho(L) \leq n_1$;

(ii) $R_\alpha(L) \geq c(s, \alpha) n_1^{-\alpha}$ *for $\alpha > 1$, where $c(s, \alpha) > 0$ depends only on s and α ;*

(iii) $R_1(L) \geq c(s) n_1^{-1} \log(N/n_1)$, *where $c(s) > 0$ depends only on s;*

(iv) $D(L) \geq 1 - (1 - n_1^{-1})^s \geq n_1^{-1}$.

Proof. (i) Since the invariants $n_2, \ldots, n_r$ are divisors of n_1, it follows from Theorem A that the coordinates of all points of L are rationals with denominator n_1. Therefore $L^\perp$ contains $n_1 \mathbf{Z}^s$. In particular, we have $\mathbf{h}_0 = (n_1, 0, \ldots, 0) \in L^\perp$, hence $\varrho(L) \leq r(\mathbf{h}_0) = n_1$.

(ii) Since $L^\perp \supseteq n_1 \mathbf{Z}^s$ by the proof of (i), we get for $\alpha > 1$:

$$R_\alpha(L) \geq \sum_{\substack{\mathbf{h} \in n_1 \mathbf{Z}^s \\ \mathbf{h} \neq 0}} r(\mathbf{h})^{-\alpha} = \sum_{\mathbf{h} \in n_1 \mathbf{Z}^s} r(\mathbf{h})^{-\alpha} - 1$$

$$= \left(\sum_{h \in \mathbf{Z}} r(n_1 h)^{-\alpha} \right)^s - 1 = \left(1 + 2 \sum_{h=1}^{\infty} (n_1 h)^{-\alpha} \right)^s - 1$$

$$= \left(1 + 2\xi(\alpha) n_1^{-\alpha} \right)^s - 1 \geq c(s, \alpha) n_1^{-\alpha},$$

where ξ is the Riemann zeta-function.

(iii) From $L^{\perp} \supseteq n_1 \mathbf{Z}^s$ we get $E(L) \supseteq C_s^*(N) \cap (n_1 \mathbf{Z}^s)$. The elements of the latter set are exactly all points $n_1 \mathbf{h}$ with $\mathbf{h} \in C_s^*(N/n_1)$. Therefore

$$R_1(L) \geq \sum_{\mathbf{h} \in C_s^*(N/n_1)} r(n_1 \mathbf{h})^{-1} = \sum_{\mathbf{h} \in C_s(N/n_1)} r(n_1 \mathbf{h})^{-1} - 1$$

$$= \left(\sum_{h \in C_1(N/n_1)} r(n_1 h)^{-1} \right)^s - 1 = \left(1 + \frac{1}{n_1} \sum_{h \in C_1^*(N/n_1)} |h|^{-1} \right)^s - 1$$

$$\geq c(s) n_1^{-1} \log(N/n_1).$$

(iv) By the proof of (i), the coordinates of all points of the node set of L are rationals with denominator n_1. Therefore, all points of the node set of L belong to the interval $J_0 = [0, 1 - n_1^{-1}]^s$. It follows that

$$D(L) \geq |\frac{1}{N} \#\{1 \leq n \leq N : \mathbf{x}_n \in J_0\} - \text{Vol}(J_0)|$$

$$= 1 - (1 - n_1^{-1})^s \geq n_1^{-1}. \quad \Box$$

Recall that an efficient lattice rule L should have a large value of $\varrho(L)$ and small values of $D(L)$ and $R_\alpha(L)$ for $\alpha \geq 1$. Thus the bounds in Theorem 1 carry the same information, namely that the first invariant n_1 must be large in order to get an efficient lattice rule. In other words, if N is fixed, then $n_2, \ldots, n_r$ should be small.

3. Existence of efficient lattice rules

In this section we discuss theorems guaranteeing the existence of lattice rules yielding small integration errors. The specific form of these existence theorems depends on the rank of the lattice rule (see Definition 2). It is convenient to prove these theorems for the quantity $R_1(L)$, since the discrepancy $D(L)$ can be bounded in terms of $R_1(L)$ according to (5), and since we have the general inequality

$$R_\alpha(L) \leq c(s, \alpha) R_1(L)^\alpha \quad \text{for} \quad \alpha > 1 \tag{6}$$

by a result of Niederreiter [17], where $c(s, \alpha) > 0$ depends only on s and α.

Lattice rules of rank 1 have already been studied for several decades under the name *method of good lattice points* (see [12] for a survey), and the following general existence theorem was shown by Niederreiter [13].

Theorem B. *For every $s \geq 2$ and $N \geq 2$ there exists an N-point lattice rule L in dimension s of rank 1 with*

$$R_1(L) = O(N^{-1}(\log N)^s).$$

This result is in fact best possible since it was proved by Larcher [9] that for any N-point lattice rule L in dimension s of rank 1, $R_1(L)$ is at least of the order of magnitude $N^{-1}(\log N)^s$. For lattice rules of rank 2 we have the following recent existence theorem of Niederreiter [17].

Theorem C. *For every $s \geq 2$ and any integers $n_1 \geq 2$ and $n_2 \geq 2$ with n_2 dividing n_1, there exists an s-dimensional lattice rule L of rank 2 with invariants n_1 and n_2 such that*

$$R_1(L) = O(N^{-1}(\log N)^s + n_1^{-1} \log N),$$

where $N = n_1 n_2$ is the number of nodes.

The implied constants in Theorems B and C depend only on s. It should be noted that the results in Theorems B and C are obtained by calculating the mean value of $R_1(L)$ over a family of lattice rules L with prescribed rank and invariants. The bounds for $R_1(L)$ in Theorems B and C are thus satisfied "on the average" for lattice rules from such a family.

For general ranks we get somewhat weaker existence theorems by comparing efficiency measures for sublattices. Let $L_1 \subseteq L_2$ be lattice rules of the same dimension s, and for $i = 1, 2$ let N_i be the number of nodes of L_i. Note that N_1 divides N_2 since $L_1 / \mathbf{Z}^s$ is a subgroup of order N_1 of the group $L_2 / \mathbf{Z}^s$ of order N_2.

Lemma 1. *For $L_1 \subseteq L_2$ and N_1, N_2 as above we have:*

(i) $R_\alpha(L_2) \leq R_\alpha(L_1)$ *for* $\alpha > 1$;

(ii) $\varrho(L_2) \geq \varrho(L_1)$;

(iii) $R_1(L_2) < \left(1 + 2 \log \frac{2N_2}{N_1}\right)^s \left(\frac{1}{N_1} + R_1(L_1)\right)$.

Proof. (i) From $L_1 \subseteq L_2$ we get $L_1^\perp \supseteq L_2^\perp$, and so the desired inequality follows immediately from the definitions of $R_\alpha(L_1)$ and $R_\alpha(L_2)$.

(ii) Use $L_1^\perp \supseteq L_2^\perp$ and the definition of the figure of merit.

(iii) Since $E(L_2) \subseteq C_s^*(N_2) \cap L_1^\perp$, we have

$$R_1(L_2) \leq \sum_{\mathbf{h} \in C_s^*(N_2) \cap L_1^\perp} r(\mathbf{h})^{-1} = R_1(L_1) + \sum_{\substack{\mathbf{h} \in C_s(N_2) \cap L_1^\perp \\ \mathbf{h} \notin C_s(N_1)}} r(\mathbf{h})^{-1}$$

$$=: R_1(L_1) + \sum .$$

Any $\mathbf{h} \in C_s(N_2)$ can be written in the form $\mathbf{h} = \mathbf{k} + N_1 \mathbf{m}$ with $\mathbf{k} \in C_s(N_1), \mathbf{m} \in I_s :=$ $[-N_2/(2N_1), N_2/(2N_1)]^s \cap \mathbf{Z}^s$. If $\mathbf{h} \notin C_s(N_1)$, then we must have $\mathbf{m} \neq \mathbf{0}$. Since $N_1 \mathbf{m} \in L_1^\perp$ by Theorem A, we have $\mathbf{h} \in L_1^\perp$ if and only if $\mathbf{k} \in L_1^\perp$. Thus with $M_s = I_s \setminus \{\mathbf{0}\}$ we get

$$\sum \leq \sum_{\mathbf{m} \in M_s} \sum_{\mathbf{k} \in C_s(N_1) \cap L_1^\perp} r(\mathbf{k} + N_1 \mathbf{m})^{-1}$$

$$= \sum_{\mathbf{m} \in M_s} r(N_1 \mathbf{m})^{-1} + \sum_{\mathbf{m} \in M_s} \sum_{\mathbf{k} \in E(L_1)} r(\mathbf{k} + N_1 \mathbf{m})^{-1}.$$

Now we claim that

$$r(\mathbf{k} + N_1 \mathbf{m}) \geq r(\mathbf{k}) r(\mathbf{m}) \quad \text{for} \quad \mathbf{k} \in C_s(N_1), \mathbf{m} \in \mathbf{Z}^s. \tag{7}$$

It suffices to show

$$r(k + N_1 m) \geq r(k) r(m) \quad \text{for} \quad k \in C_1(N_1), m \in \mathbf{Z}.$$

This inequality is trivial whenever $km = 0$. If $km \neq 0$, then

$$r(k + N_1 m) = |k + N_1 m| \geq N_1 |m| - |k| \geq N_1 |m| - \frac{1}{2} N_1$$

$$= \frac{1}{2} N_1 (2|m| - 1) \geq |k||m| = r(k) r(m).$$

Thus (7) is proved. Using (7) we get

$$\sum \leq \frac{1}{N_1} \sum_{\mathbf{m} \in M_s} r(\mathbf{m})^{-1} + \sum_{\mathbf{m} \in M_s} r(\mathbf{m})^{-1} \sum_{\mathbf{k} \in E(L_1)} r(\mathbf{k})^{-1}$$

$$= \left(\sum_{\mathbf{m} \in M_s} r(\mathbf{m})^{-1} \right) \left(\frac{1}{N_1} + R_1(L_1) \right)$$

$$= \left(\sum_{\mathbf{m} \in I_s} r(\mathbf{m})^{-1} - 1 \right) \left(\frac{1}{N_1} + R_1(L_1) \right).$$

Therefore

$$R_1(L_2) < \left(\sum_{\mathbf{m} \in I_s} r(\mathbf{m})^{-1} \right) \left(\frac{1}{N_1} + R_1(L_1) \right).$$

Now

$$\sum_{\mathbf{m} \in I_s} r(\mathbf{m})^{-1} = \left(1 + 2 \sum_{m=1}^{\lfloor N_2/2N_1 \rfloor} m^{-1} \right)^s.$$

As in [11, Lemma 3.7] one shows that

$$\sum_{m=1}^{\lfloor M/2 \rfloor} m^{-1} < \log(2M) \quad \text{for any integer} \quad M \geq 1,$$

and this yields the result of (iii). $\Box$

Theorem 2. *For a given dimension $s \geq 2$ let a rank r and invariants $n_1, \ldots, n_r$ satisfying the conditions in Theorem A be prescribed. Then there exists an s-dimensional lattice rule L with this rank and these invariants which has the following properties:*

$$R_1(L) \leq \frac{c(s)}{n_1} (\log n_1)^s \left(1 + \log \frac{N}{n_1} \right)^s,$$

$$R_\alpha(L) \leq c(s, \alpha) n_1^{-\alpha} (\log n_1)^{s\alpha} \quad \text{for} \quad \alpha > 1,$$

where $N = n_1 \ldots n_r$ is the number of nodes of L and the constants $c(s)$ and $c(s, \alpha)$ depend only on the indicated parameters.

Proof. Let L_1 be an n_1-point lattice rule in dimension s of rank 1 which satisfies $R_1(L_1) = O(n_1^{-1}(\log n_1)^s)$ with an implied constant depending only on s. Such a lattice rule exists according to Theorem B. Then let $L \supseteq L_1$ be an s-dimensional lattice rule of rank r with invariants $n_1, \ldots, n_r$. Group-theoretically this means that we extend a cyclic group of order n_1 by a direct sum of cyclic groups of orders $n_2, \ldots, n_r$, respectively (compare with Sloan and Lyness [24]). Lemma 1 (iii) yields the desired bound for $R_1(L)$. For $\alpha > 1$ we apply (6) to get

$$R_\alpha(L_1) \leq c(s, \alpha) n_1^{-\alpha} (\log n_1)^{s\alpha},$$

and then an application of Lemma 1 (i) completes the proof. $\Box$

Theorem 3. *For a given dimension $s \geq 2$ let a rank r and invariants $n_1, \ldots, n_r$ satisfying the conditions in Theorem A be prescribed. Then there exists an s-dimensional lattice rule L with this rank and these invariants for which*

$$\varrho(L) \geq c(s) n_1^{-1} (\log n_1)^{1-s},$$

where the constant $c(s) > 0$ depends only on s.

Proof. Let L_1 be an n_1-point lattice rule in dimension s of rank 1 for which

$$\varrho(L_1) \geq c(s)n_1^{-1}(\log n_1)^{1-s}.$$

Such a lattice rule exists by a theorem of Zaremba [27]. Now proceed as in the proof of Theorem 2 and use Lemma 1 (ii).□

It should be pointed out that the proofs of the existence theorems presented so far in this section are nonconstructive. Explicit efficient lattice rules have been obtained by computer searches. For efficient lattice rules of rank 1 we refer to the extensive tables in Maisonneuve [10] and Hua and Wang [6]. Sloan and Walsh [25] carried out wide-ranging searches and found many lattice rules of rank 2 which perform better than lattice rules of rank 1 with a comparable number of nodes. All these efficient lattice rules of rank 2 have a small second invariant n_2 (usually $n_2 = 2$), in accordance with our remark following Theorem 1. For the maximal rank $r = s$, Disney and Sloan [1] obtained lattice rules that also improve on lattice rules of rank 1; these lattice rules of rank s have invariants $2m, 2, \ldots, 2$ with m being a sufficiently large integer.

4. Generalization to compact groups

The natural setting for a general theory of lattice rules is a compact group. In the case considered above the compact group is $\mathbf{R}^s/\mathbf{Z}^s$ and a lattice rule is defined in terms of a finite subgroup of this group. Now let G be an arbitrary compact group, which we assume to be Hausdorff. Let μ be the uniquely determined Haar measure on G, i.e. a translation-invariant regular Borel probability measure on G (in the case $G = \mathbf{R}^s/\mathbf{Z}^s$ this is the measure induced by the s-dimensional Lebesgue measure). For the necessary background on compact groups we refer to Hewitt and Ross [4] and Pontryagin [20] and to the synopsis in Kuipers and Niederreiter [8, Ch. 4].

Definition 3. Let H be a finite subgroup of order $|H|$ of the compact group G. Then the *finite subgroup rule* corresponding to H uses the approximation

$$\int\limits_G f d\mu \approx \frac{1}{|H|} \sum_{x \in H} f(x) \quad \text{for} \quad f \in L^1(G).$$

For the error analysis we fix G and H and we use the representation theory of G. From each equivalence class of irreducible representations of G we choose one representative, thus obtaining a system $\{D^{(\lambda)} : \lambda \in \wedge\}$ of nonequivalent irreducible representations of G, where $\wedge$ is a suitable index set. We arrange the notation in such a way that $0 \in \wedge$ and that $D^{(0)}$ is the trivial representation of G, which has degree 1. Furthermore, we choose the representatives $D^{(\lambda)}$ from the equivalence classes in such a way that the following condition is satisfied: if $D^{(\lambda)}$ is considered as a representation of H, then $D^{(\lambda)}$ has a block diagonal form in which the diagonal blocks are irreducible representations of H and the trivial representations of H occurring among these diagonal blocks are listed first (compare with [8, p. 226]). For each $\lambda \in \wedge$ let $t(\lambda)$ be the number of trivial representations of H occurring in the block diagonal form of $D^{(\lambda)}$ as a representation of H. Then $0 \leq t(\lambda) \leq d(\lambda)$, where $d(\lambda)$ is the degree of $D^{(\lambda)}$. Let $D^{(\lambda)}(x)$ be the square matrix $(d_{ij}^{(\lambda)}(x))$ of order $d(\lambda)$ with $x \in G$. If x is viewed as a variable, then the entry $d_{ij}^{(\lambda)}$ belongs to $C(G)$, the space of complex-valued continuous functions on G.

If $f \in C(G)$ and $\varepsilon > 0$ are given, then by the Peter–Weyl theorem (see [8, p. 226]) there exists a $g_\varepsilon \in C(G)$ such that in the supremum norm $\|f - g_\varepsilon\| \leq \varepsilon$ and g_ε is of the form

$$g_\varepsilon = \sum_{\lambda \in \Lambda_0} \sum_{i,j=1}^{d(\lambda)} a_{ij}^{(\lambda)} d_{ij}^{(\lambda)},$$

where Λ_0 is a finite subset of Λ with $0 \in \Lambda_0$ and $a_{ij}^{(\lambda)} \in \mathbf{C}$.

Theorem 4. *For $f \in C(G)$ and $\varepsilon > 0$ let g_ε be as above. Then*

$$\left| \frac{1}{|H|} \sum_{x \in H} f(x) - \int_G f d\mu \right| \leq 2\varepsilon + \left| \sum_{\lambda \in \Lambda_0^*} \sum_{i=1}^{t(\lambda)} a_{ii}^{(\lambda)} \right|,$$

where $\Lambda_0^* = \Lambda_0 \setminus \{0\}$.

Proof. From $\|f - g_\varepsilon\| \leq \varepsilon$ we get

$$\left| \frac{1}{|H|} \sum_{x \in H} f(x) - \int_G f d\mu \right| \leq 2\varepsilon + \left| \frac{1}{|H|} \sum_{x \in H} g_\varepsilon(x) - \int_G g_\varepsilon d\mu \right|. \tag{8}$$

By the orthogonality relations for compact groups we have $\int_G d_{ij}^{(\lambda)} d\mu = 0$ for all $d_{ij}^{(\lambda)}$ with $\lambda \neq 0$, thus

$$\frac{1}{|H|} \sum_{x \in H} g_\varepsilon(x) - \int_G g_\varepsilon d\mu =$$

$$= \frac{1}{|H|} \sum_{\lambda \in \Lambda_0} \sum_{i,j=1}^{d(\lambda)} a_{ij}^{(\lambda)} \sum_{x \in H} d_{ij}^{(\lambda)}(x) - \sum_{\lambda \in \Lambda_0} \sum_{i,j=1}^{d(\lambda)} a_{ij}^{(\lambda)} \int_G d_{ij}^{(\lambda)} d\mu$$

$$= \frac{1}{|H|} \sum_{\lambda \in \Lambda_0} \sum_{i,j=1}^{d(\lambda)} a_{ij}^{(\lambda)} \sum_{x \in H} d_{ij}^{(\lambda)}(x) - a_{11}^{(0)}.$$

Therefore

$$\frac{1}{|H|} \sum_{x \in H} g_\varepsilon(x) - \int_G g_\varepsilon d\mu = \frac{1}{|H|} \sum_{\lambda \in \Lambda_0^*} \sum_{i,j=1}^{d(\lambda)} a_{ij}^{(\lambda)} \sum_{x \in H} d_{ij}^{(\lambda)}(x). \tag{9}$$

For fixed $\lambda \neq 0$ we put

$$S^{(\lambda)} = \sum_{x \in H} D^{(\lambda)}(x). \tag{10}$$

By the condition on $D^{(\lambda)}$, the matrix $S^{(\lambda)}$ has a block diagonal form with diagonal blocks $S_j^{(\lambda)} := \sum_{x \in H} E_j(x), 1 \leq j \leq m(\lambda)$, where the E_j with $1 \leq j \leq t(\lambda)$ are trivial representations of H and the E_j with $t(\lambda) < j \leq m(\lambda)$ are nontrivial irreducible representations of H. It is clear that

$$S_j^{(\lambda)} = |H| \quad \text{for} \quad 1 \leq j \leq t(\lambda).$$

For $t(\lambda) < j \leq m(\lambda)$ and any $h \in H$ we have

$$S_j^{(\lambda)} = \sum_{x \in H} E_j(hx) = E_j(h) S_j^{(\lambda)}. \tag{11}$$

If E_j has degree 1, then $E_j(h) \neq 1$ for some $h \in H$, and it follows that $S_j^{(\lambda)} = 0$. If E_j has degree > 1 and we had $S_j^{(\lambda)} \neq 0$, then $S_j^{(\lambda)} \mathbf{z} \neq \mathbf{0}$ for some complex vector $\mathbf{z}$. It follows then from (11) that with $\mathbf{y} = S_j^{(\lambda)} \mathbf{z}$ we get $E_j(h)\mathbf{y} = \mathbf{y}$ for all $h \in H$, a contradiction to the irreducibility of E_j. This shows that $S_j^{(\lambda)} = 0$ for $t(\lambda) < j \leq m(\lambda)$, where 0 denotes a zero matrix of appropriate order. By comparing entries in (10), we obtain then

$$\sum_{x \in H} d_{ij}^{(\lambda)}(x) = \begin{cases} |H| & \text{if } i = j \quad \text{and} \quad 1 \leq i \leq t(\lambda), \\ 0 & \text{otherwise.} \end{cases} \tag{12}$$

Together with (8) and (9) this shows the theorem. $\square$

Theorem 5. *If $f \in C(G)$ is represented by the uniformly and absolutely convergent series*

$$f = \sum_{\lambda \in \Lambda_0} \sum_{i,j=1}^{d(\lambda)} a_{ij}^{(\lambda)} d_{ij}^{(\lambda)},$$

where Λ_0 is a countable subset of Λ with $0 \in \Lambda_0$ and $a_{ij}^{(\lambda)} \in \mathbf{C}$, then

$$\frac{1}{|H|} \sum_{x \in H} f(x) - \int_G f \, d\mu = \sum_{\lambda \in \Lambda_0^*} \sum_{i=1}^{t(\lambda)} a_{ii}^{(\lambda)},$$

where $\Lambda_0^* = \Lambda_0 \setminus \{0\}$.

Proof. We can proceed in formal analogy with the argument leading to (9), but with g_ϵ replaced by f. Then an application of (12) yields the desired result. $\square$

If G is abelian, then its representation theory is much simpler since every irreducible representation of G has degree 1. Thus we have $d(\lambda) = 1$ for all $\lambda \in \Lambda$ and $\chi^{(\lambda)} := d_{11}^{(\lambda)}$ is a character of G. In particular, we have $t(\lambda) = 0$ or 1, and $t(\lambda) = 1$ if and only if $\chi^{(\lambda)}$ is trivial on H; these characters $\chi^{(\lambda)}$ make up the annihilator of H in the character group of G (compare with [8, p. 232]). Consequently, we get simplifications of the statements of Theorems 4 and 5 for abelian G. For instance, the integration error in Theorem 5 is now simply a sum of Fourier coefficients $a_{11}^{(\lambda)}$ of f, where λ runs through all those elements of Λ_0^* for which $\chi^{(\lambda)}$ is trivial on H.

The further development of the theory of finite subgroup rules, parallel to that for the special case of lattice rules, depends on the specific nature of the irreducible representations of G. Compact groups that are of particular interest for applications are spheres of suitable dimensions and the various compact groups of mathematical physics. Finite subgroup rules are of interest even in the case of a finite group G (where the Haar integral reduces to an arithmetic mean of function values), namely when G is of large order and a well-chosen finite subgroup rule allows a good approximation by summing over a much smaller subgroup H. Particular instances of this which are of practical relevance are groups G that are direct sums $\bigoplus_{i=1}^{s}(\mathbf{Z}/m_i\mathbf{Z})$, where $m_1, \ldots, m_s$ are integers ≥ 2. Such groups appear e.g. in the calculation of multidimensional discrete Fourier transforms. The case of finite groups could also be of interest for discrete event simulation.

5. Conclusions

We summarize some of the attractive features of lattice rules for numerical integration over $U^s = [0, 1]^s$. First of all, lattice rules are easy to implement since their node sets can be generated quickly (see Theorem A). Furthermore, the node sets of well-chosen lattice rules yield low-discrepancy point sets. The results of Section 3 show that for every dimension $s \geq 2$ and every $N \geq 2$ there exist s-dimensional N-point lattice rules L with discrepancy $D(L) = O(N^{-1}(\log N)^s)$. In view of (2), such a lattice rule involves then an error bound of the form $O(N^{-1}(\log N)^s)$ for integrands that are of bounded variation on U^s in the sense of Hardy and Krause. If the integrand is periodic with U^s as period interval and belongs to the regularity class $\mathcal{E}_\alpha^s$ (see Definition 1), then even smaller error bounds are available. For instance, (4), (6), and Theorem B imply that for such integrands we get an error bound of the form $O(N^{-\alpha}(\log N)^{s\alpha})$ with a suitably chosen lattice rule. A wide variety of concrete lattice rules achieving these error bounds is available (see Section 3).

The generalization of lattice rules discussed in Section 4 leads to promising integration rules for compact groups. Further studies for specific compact groups have to be carried out to devise efficient finite subgroup rules in a concrete form.

References

1. S. Disney and I.H. Sloan: Lattice integration rules of maximal rank, preprint, University of New South Wales, Sydney, 1990.

2. H. Faure: Discrépance de suites associées à un système de numération (en dimension s), *Acta Arith.* **41**, 337-351 (1982).

3. K.K. Frolov: On the connection between quadrature formulas and sublattices of the lattice of integral vectors (Russian), *Dokl. Akad. Nauk SSSR* **232**, 40-43 (1977).

4. E. Hewitt and K.A. Ross: *Abstract Harmonic Analysis I, II*, Springer, Berlin, 1963, 1970.

5. E. Hlawka: Zur angenäherten Berechnung mehrfacher Integrale, *Monatsh. Math.* **66**, 140-151 (1962).

6. L.K. Hua and Y. Wang: *Applications of Number Theory to Numerical Analysis*, Springer, Berlin, 1981.

7. N.M. Korobov: The approximate computation of multiple integrals (Russian), *Dokl. Akad. Nauk SSSR* **124**, 1207-1210 (1959).

8. L. Kuipers and H. Niederreiter: *Uniform Distribution of Sequences*, Wiley, New York, 1974.

9. G. Larcher: A best lower bound for good lattice points, *Monatsh. Math.* **104**, 45-51 (1987).

10. D. Maisonneuve: Recherche et utilisation des "bons treillis". Programmation et résultats numériques, *Applications of Number Theory to Numerical Analysis* (S.K. Zaremba, ed.), pp.121-201, Academic Press, New York, 1972.

11. H. Niederreiter: Pseudo-random numbers and optimal coefficients, *Advances in Math.* **26**, 99-181 (1977).

12. H. Niederreiter: Quasi-Monte Carlo methods and pseudo-random numbers, *Bull. Amer. Math. Soc.* **84**, 957-1041 (1978).

13. H. Niederreiter: Existence of good lattice points in the sense of Hlawka, *Monatsh. Math.* **86**, 203-219 (1978).

14. H. Niederreiter: Point sets and sequences with small discrepancy, *Monatsh. Math.* **104**, 273-337 (1987).

15. H. Niederréiter: Quasi-Monte Carlo methods for multidimensional numerical integration, *Numerical Integration III* (H. Braß and G. Hämmerlin, eds.), pp. 157-171, Internat. Series of Numerical Math., Vol. 85, Birkhäuser, Basel, 1988.

16. H. Niederreiter: Low-discrepancy and low-dispersion sequences, *J. Number Theory* **30**, 51-70 (1988).

17. H. Niederreiter: The existence of efficient lattice rules for multidimensional numerical integration, preprint, Austrian Academy of Sciences, Vienna, 1990.

18. H. Niederreiter: *Random Number Generation and Quasi-Monte Carlo Methods*, SIAM, Philadelphia, in preparation.

19. H. Niederreiter and I.H. Sloan: Lattice rules for multiple integration and discrepancy, *Math. Comp.* **54**, 303-312 (1990).

20. L.S. Pontryagin: *Topological Groups*, 2nd ed., Gordon and Breach, New York, 1966.

21. I.H. Sloan: Lattice methods for multiple integration, *J. Comput. Appl. Math.* **12/13**, 131-143 (1985).

22. I.H. Sloan and P.J. Kachoyan: Lattices for multiple integration, *Mathematical Programming and Numerical Analysis Workshop* (Canberra, 1983), pp. 147-165, Proc. Centre Math. Anal. Austral. Nat. Univ., Vol. 6, Austral. Nat. Univ., Canberra, 1984.

23. I.H. Sloan and P.J. Kachoyan: Lattice methods for multiple integration: theory, error analysis and examples, *SIAM J. Numer. Anal.* **24**, 116-128 (1987).

24. I.H. Sloan and J.N. Lyness: The representation of lattice quadrature rules as multiple sums, *Math. Comp.* **52**, 81-94 (1989).

25. I.H. Sloan and L. Walsh: A computer search of rank-2 lattice rules for multidimensional quadrature, *Math. Comp.* **54**, 281-302 (1990).

26. I.M. Sobol': The distribution of points in a cube and the approximate evaluation of integrals (Russian), *Zh. Vychisl. Mat. i Mat. Fiz.* **7**, 784-802 (1967).

27. S.K. Zaremba: Good lattice points modulo composite numbers, *Monatsh. Math.* **78**, 446-460 (1974).

Limit theorems on the Robbins-Monro process
for different variance behaviors of the
stochastic gradient

Ernst Plöchinger

Universität der Bundeswehr München

Fakultät für Luft- und Raumfahrttechnik

Werner-Heisenberg-Weg 39

D(W)-8014 Neubiberg

Abstract

For finding a solution $x^* \in \mathbb{R}^r$ of the system of nonlinear equations $G(x)=0$ in case that only estimations $\hat{G}(x) = G(x)+Z(x)$ at each $x \in \mathbb{R}^r$ are available, we consider the stochastic approximation procedure

$$X_{n+1} := X_n - r_n \hat{G}_n(X_n), \text{ where } \hat{G}_n(X_n) := G(X_n)+Z_n, \quad n=1,2,\ldots, \qquad (1)$$

with the estimation error $Z_n := Z(X_n)$. In this paper the limiting distribution of the random sequence $(r_n(X_n-x^*))_n$ is considered for different sequences $(r_n)_n$ of deterministic step sizes, where $(r_n)_n$ is a sequence of positive numbers such that

$$E||X_n-x^*||^2 = O(1/r_n^2).$$

Approximations for the limit covariance matrices of $(r_n(X_n-x^*))_n$ are given for the case where the estimation error Z_n in (1) has **different variances** for indices n from different subsets $\mathbb{N}^{(k)}$ of $\mathbb{N}$. Particular attention is paid to the semi-stochastic case where $Z_n=0$ for all n contained in an infinite subset $N^{(1)}$ of $\mathbb{N}$.

1. Introduction

For a given measurable function $G: \mathbb{R}^{\nu} \to \mathbb{R}^{\nu}$ we want to determine $x^* \in \mathbb{R}^{\nu}$, where

$$G(x^*) = 0. \tag{1}$$

In order to determine x^* according to (1), the following stochastic approximation method (Robbins-Monro process) will be applied:

Select $X_1 \in \mathbb{R}^{\nu}$, and for $n = 1, 2, \ldots$ compute (2.1)

$$X_{n+1} := X_n - r_n \cdot \hat{G}_n(X_n), \text{ where } \hat{G}_n(X_n) = G(X_n) + Z_n. \tag{2.2}$$

$(r_n)_n$ is the sequence of positive __step sizes__, and $(Z_n)_n$ is a sequence of random vectors in a probability space $(\Omega, \mathcal{O}\!\!\!\!\textit{l}, P)$.

Z_n is the __error of estimation__ made at step n in the process of calculating $G(X_n)$.

Reference [12] was the first to indicate conditions of the one-dimensional case ($\nu = 1$) where the sequence of random vectors $(X_n)_n$ derived from (2) converges - in mean square - to $x^* \in \mathbb{R}^{\nu}$ according to (1).

Since then, numerous papers on issues of convergence according to (2) have been published. One of the first surveys of known convergence statements is to be found in [14].

[3] shows that under suitable conditions, and using the classical (standard) step sitze $r_n = \frac{M}{n}$ (M>o), the following equation applies to the mean square error:

$$E||X_n - x^*||^2 = 0(\frac{1}{r_n^2}).$$

In this case, according to [13] und [4], the sequence $(\sqrt{n}(X_n - x^*))_n$ - given further requirements - is characterized by an asymptotic N(o,V)-distribution with the covariance matrix V being calculated from the data on G(x) and $(Z_n)_n$. This paper develops theorems on the limiting distribution of the sequence $(r_n \cdot (X_n - x^*))_n$ using different deterministic step sizes r_n in equation (2), where

(r_n) is a sequence of positive numbers such that

$$E||X_n-x^*||^2 = O(\frac{1}{r_n^2}).$$

Approximations for the limit covariance matrices of $(r_n(X_n-x^*))_n$ are given for the case where the estimation error Z_n in (2.2) has different variances for indices n from different subsets $\mathbb{N}^{(k)}$ of $\mathbb{N}$. Particular attention is paid to the semi-stochastic case where $Z_n=0$ for all n contained in an infinite subset $\mathbb{N}^{(1)}$ of $\mathbb{N}$.

The results will subsequently be applied to a situation where the sequence of estimation errors $(Z_n)_n$ is represented by a special approach.

1.1. General premises and notations

In order to obtain convergence in (2), we start from the following valid premises applying to the function G and the sequence of estimation errors $(Z_n)_n$:

a) Assume $x^* \in \mathbb{R}^\nu$, positive numbers α and β, a symmetric matrix H and a function $\delta: \mathbb{R}^\nu \rightarrow \mathbb{R}^\nu$, where

$$\alpha||x-x^*||^2 \leq <G(x),x-x^*>, \tag{3.1}$$

$$||G(x)|| \leq \beta ||x-x^*||, \tag{3.2}$$

$$G(x) = H \cdot (x-x^*) + \delta(x) \text{ and} \tag{3.3}$$

$$\delta(x) = o(||x-x^*||) \tag{3.4}$$

for all $x \in \mathbb{R}^\nu$.

b) Assume numbers $\sigma, \gamma \geq 0$ such that for all $n=1,2,\ldots$

$$E(Z_n|\mathcal{O}_n) \equiv 0, \tag{4.1}$$

$$E(||Z_n||^2) \leq \sigma^2 + \gamma^2 \cdot E||X_n-x^*||^2, \tag{4.2}$$

where $\mathcal{O}_n$ is the σ-algebra generated by $X_1,\ldots,X_n,Z_1,\ldots,Z_{n-1}$. Equations (3.3) and (3.4) signify that G is differentiable in x^* and has the Jacobian matrix H at x^*. Furthermore, (3) yields for each eigenvalue h of H

$$\alpha \le h \le \beta. \tag{5}$$

According to (4.1), $\hat{G}_n(X_n) = G(X_n) + Z_n$ is an "unbiased" estimate of $G(X_n)$ for a given α_n, i.e., the following holds true

$$E(\hat{G}_n(X_n) | \alpha_n) = G(X_n).$$

Let P be an orthogonal matrix with

$$P^T H P = \begin{pmatrix} h_1 & & 0 \\ & \ddots & \\ 0 & & h_\nu \end{pmatrix}, \tag{6}$$

where $h_1, \ldots, h_\nu$ are the eigenvalues of H.

With respect to indices $m, n \in N$ with $m \le n$ take

$$B_{m,n} := \begin{cases} \prod\limits_{j=m+1}^{n} (I - r_j H) & , \text{ for } m < n \\ I & , \text{ for } m = n \end{cases}$$

and for $i \in (1, \ldots, \nu)$

$$B_{m,n}^{(i)} := \begin{cases} \prod\limits_{j=m+1}^{n} (1 - r_j h_i) & , \text{ for } m < n \\ 1 & , \text{ for } m = n \end{cases}$$

Because of (6) we obtain

$$P^T B_{m,n} P = \begin{pmatrix} B_{m,n}^{(1)} & & 0 \\ & \ddots & \\ 0 & & B_{m,n}^{(\nu)} \end{pmatrix} =: \hat{B}_{m,n}. \tag{7}$$

For abbreviation purpose let

$$\Delta_n := X_n - x^* \quad , n = 1, 2, \ldots$$

be the <u>iteration error</u> in the n-th step, and for a subset $M \subseteq N$ and $m \le n$ let

$$M_{m,n} := (m+1, \ldots, n) \cap M.$$

be the <u>(m,n)-segment</u> of M.

2. <u>Important auxiliary propositions</u>

The following auxiliary proposition is important for proving the convergence of the mean square errors $(E||\Delta_n||^2)_n$:

Lemma 2.1. For all $n \in \mathbb{N}$ the following holds true:

$$E(||\Delta_{n+1}||^2 |\alpha_n) \leq r_n^2 E(||Z_n||^2 |\alpha_n) + (1-2\alpha r_n + \beta^2 r_n^2) \cdot ||\Delta_n||^2.$$

Proof. (cf. [8]). Because of (2.2) the following holds true for any $n \in \mathbb{N}$:

$$||\Delta_{n+1}||^2 = ||\Delta_n||^2 + r_n^2 ||G(X_n)+Z_n||^2 - 2r_n <\Delta_n, G(X_n)+Z_n>.$$

Relations (4.1),(3.1) and (3.2) then yield

$$E(||\Delta_{n+1}||^2 |\alpha_n) \leq ||\Delta_n||^2 + r_n^2 (\beta^2 ||\Delta_n||^2 + E(||Z_n||^2 |\alpha_n))$$
$$- 2r_n \alpha ||\Delta_n||^2.$$

An explicit formula for the error Δ_{n+1} in the (n+1)-th step can be specified using the matrices $B_{m,n}$ given in section 1.1:

Lemma 2.2. For all $n \in \mathbb{N}$ the following holds true:

$$\Delta_{n+1} = (I - r_n H) \cdot \Delta_n - r_n (\delta(X_n) + Z_n)$$
$$= B_{o,n} \Delta_1 - \sum_{m=1}^{n} B_{m,n} r_m (\delta(X_m) + Z_m).$$

Proof. Because of (2) und (3.3) the following holds true for all $n \in \mathbb{N}$:

$$\Delta_{n+1} = X_n - r_n \cdot ((H \cdot (X_n - x^*) + \delta(X_n)) + Z_n) - x^*.$$

Hence, the assertion of this lemma follows by complete induction on n.

According to Lemma 2.2 $r_n \cdot \Delta_{n+1}$ may be specified as the sum of three terms. One criterion verifying that the first two terms converge to 0 is given by

Lemma 2.3. If r_n, r_n, Δ_n and $B_{m,n}$ fulfill the following assumptions

a) $\lim_{n \to \infty} r_n \cdot ||B_{m,n}|| = 0$ for all $m \in \mathbb{N}_0$,

b) there is a constant L with

$$r_n \cdot \sum_{m=1}^{n} ||B_{m,n}|| r_m E ||\Delta_m|| \leq L$$

for all $n \in \mathbb{N}$,

c) $\lim_{n \to \infty} ||\Delta_n|| = 0$ almost sure (a.s.),

then the sequence

$$(\tau_n \cdot (B_{o,n} \Delta_1 - \sum_{m=1}^{n} B_{m,n} \cdot r_m \delta(X_m)))_n$$

stochastically converges to 0.

<u>Proof.</u> (according to [13]). Due to assumption a) we merely have to prove that the sequence

$$U_n: = \tau_n \sum_{m=1}^{n} ||B_{m,n}|| \, r_m \, ||\delta(X_m)||, \quad n=1,2,\ldots$$

stochastically converges to 0. Let ϵ and δ be arbitrary positive numbers. Because of (3.4) there is a positive number t for $s: = \min(\epsilon, \frac{\delta}{L+1})$ such that

$$||\delta(x)|| \leq \frac{s^2}{4} \, ||x-x^*|| \tag{i}$$

for all $x \in \mathbb{R}^\nu$ with $||x-x^*|| \leq t$.

For $a_{m,n}: = \tau_n \, ||B_{m,n}|| \, r_m$ we obtain

$$\mathbb{P}(U_n > \epsilon) \leq \mathbb{P}(U_n > s) \leq \tag{ii}$$

$$\mathbb{P}(\max_{m \leq k}||\delta(x_m)|| \sum_{m=1}^{k} a_{m,n} > \frac{s}{2}) + \mathbb{P}(\sum_{m=k+1}^{n} a_{m,n}||\delta(X_m)|| > \frac{s}{2}),$$

$$\mathbb{P}(\sum_{m=k+1}^{n} a_{m,n}||\delta(X_m)|| > \frac{s}{2}) \leq \mathbb{P}(\bigcup_{m>k} (||\Delta_m|| > t)) +$$

$$+ \mathbb{P}((\sum_{m=k+1}^{n} a_{m,n}||\delta(X_m)|| > \frac{s}{2}) \bigcap_{m>k} (||\Delta_m|| \leq t)), \tag{iii}$$

and because of (i)

$$\mathbb{P}((\sum_{m=k+1}^{n} a_{m,n}||\delta(X_m)|| > \frac{s}{2}) \bigcap_{m>k} (||\Delta_m|| \leq t)) \leq$$

$$\mathbb{P}(\frac{s^2}{4} \sum_{m=k+1}^{n} a_{m,n}||\Delta_m|| > \frac{s}{2}) \leq \frac{s}{2} \sum_{m=1}^{n} a_{m,n} E||\Delta_m|| \tag{iv}$$

for all $k,n \in \mathbb{N}$ with $k \leq n$.

According to assumption c) there is a $k_o \in N$ with

$$P(\bigcup_{m \geq k_o} (||\Delta_m|| > t)) \leq \frac{s}{2} .$$

Hence, because of (ii), (iii), (iv) and assumption b) we find

$$\mathbb{P}(U_n > \epsilon) \leq \mathbb{P}(\max_{m \leq k_o} ||\delta(X_m)|| \sum_{m=1}^{k_o} a_{m,n} > \frac{s}{2}) + \frac{s}{2}(L+1). \qquad (v)$$

Furthermore, there is a constant R≥o with

$$\mathbb{P}(\max_{m \leq k_o} ||\delta(X_m)|| > R) \leq \frac{\delta}{2},$$

and because of assumption a) there is an index $n_o > k_o$ with

$$\sum_{m=1}^{k_o} a_{m,n} \leq \frac{s}{2R} \text{ for all } n \geq n_o.$$

Hence, for all $n \geq n_o$ we obtain

$$\mathbb{P}(\max_{m \leq k_o} ||\delta(X_m)|| \sum_{m=1}^{k_o} a_{m,n} > \frac{s}{2}) \leq \mathbb{P}(\max_{m \leq k_o} ||\delta(X_m)|| \frac{s}{2R} > \frac{s}{2}) \leq \frac{\delta}{2}. \qquad (vi)$$

(v) and (vi) finally yield

$$\mathbb{P}(U_n > \epsilon) \leq \frac{\delta}{2} + \frac{s}{2}(L+1) \leq \delta$$

for all $n \geq n_o$.

3. Modified standard step size

In this section we start from a given disjoint decomposition

$$\mathbb{N} = \mathbb{N}^{(1)} \cup \ldots \cup \mathbb{N}^{(\kappa)} \qquad (8.1)$$

of $\mathbb{N}$ such that for all $k \in \{1, \ldots, \kappa\}$ there is the limit

$$q_k := \lim_{n \to \infty} \frac{|\{1, \ldots, n\} \cap \mathbb{N}^{(k)}|}{n} \qquad (8.2)$$

which is positive. We will consider the following step size

$$r_n := \frac{M_n}{n} \quad , \quad n = 1, 2, \ldots, \qquad (9.1)$$

where $(M_n)_n$ is a positive sequence with positive limits

$$M^{(k)} := \lim_{\mathbb{N}^{(k)} \ni m \to \infty} M_m \qquad (9.2)$$

for $k = 1, \ldots, \kappa$.

For α according to (3.1) and for

$$\tilde{M} := q_1 M^{(1)} + \ldots + q_\kappa M^{(\kappa)}$$

we assume

$$\bar{M} \alpha > \frac{1}{2} . \tag{9.3}$$

This step size will satisfy the conditions

$$\sum_{n \geq 1} r_n = \infty \quad \text{and} \quad \sum_{n \geq 1} r_n^2 < \infty.$$

According to [2] the sequence $(X_n)_n$ given in (2) converges almost sure to x^*, and, according to [3], $E||\Delta_n||^2 = 0(\frac{1}{n})$.

For the sake of completeness we will derive again these propositions in the following.

Theorem 3.1 The following holds true

a) $E||\Delta_n||^2 = 0(\frac{1}{n})$

b) $\lim_{n \to \infty} ||\Delta_n|| = 0$ almost sure.

Proof. According to Lemma 2.1 und (4.2) we find for all $n \in \mathbb{N}$

$$E||\Delta_{n+1}||^2 \leq (1-2\alpha r_n + (\beta^2+\gamma^2)r_n^2)E||\Delta_n||^2 + r_n^2\sigma^2.$$

Because of (9.3) a number h may be selected such that

$$\frac{1}{\bar{M}} < h < 2\alpha. \tag{i}$$

Since $(r_n)_n$ is a null sequence, there is an index n_0 such that for all $n \geq n_0$

$$E||\Delta_{n+1}||^2 \leq (1 - h \cdot r_n) E||\Delta_n||^2 + r_n^2\sigma^2.$$

By complete induction we obtain

$$E||\Delta_{n+1}||^2 \leq b_{n_0,n} E||\Delta_{n_0+1}||^2 + (\sigma \hat{M})^2 \sum_{m=n_0+1}^{n} b_{m,n} \frac{1}{m^2} \tag{ii}$$

for all $n \geq n_0$, where

$$\hat{M}: = \sup_n M_n \quad \text{and} \quad b_{m,n}: = \prod_{j=m+1}^{n} (1 - \frac{h \cdot M_j}{j}) .$$

(i) yields $A: = h \cdot \hat{M} - 1 > 0$, hence, because of (ii) and Lemma A.7 with $t:=1$ we obtain

$$\overline{\lim_{n \to \infty}} \; n \cdot E||\Delta_{n+1}||^2 \leq (\sigma \cdot \hat{M})^2 \frac{1}{A} < \infty.$$

In order to prove b) we define for $n \in \mathbb{N}$

$$U_n: = \sum_{k=n}^{\infty} r_k^2 E(||Z_k||^2 | \alpha_n)$$

$$V_n: = ||\Delta_n||^2 + U_n.$$

(4.2) yields for all $n \in \mathbb{N}$

$$0 \leq E U_n \leq E V_n \leq E||\Delta_n||^2 + \sum_{k \geq n} r_k^2 \cdot (\sigma^2 + \gamma^2 E||\Delta_k||^2)$$

$$\leq \sup_{k \in \mathbb{N}} E||\Delta_k||^2 \cdot (1 + \gamma^2 \sum_{k \in \mathbb{N}} r_k^2) + \sigma^2 \sum_{k \in \mathbb{N}} r_k^2 =: L,$$

where $L < \infty$ according to proposition a) and (9). Hence

$$0 \leq \sup_{n \in \mathbb{N}} E U_n \leq \sup_{n \in \mathbb{N}} E V_n < \infty. \tag{iii}$$

Since for all $n \in \mathbb{N}$ $\alpha_n \subseteq \alpha_{n+1}$, we find

$$E(U_{n+1} | \alpha_n) - \sum_{k=n+1}^{\infty} r_k^2 E(||Z_k||^2 | \alpha_n) \leq U_n, \tag{iv}$$

thus it follows that $(U_{n+1}, \alpha_n)_{n \geq 1}$ is a supermartingale.

The equation contained in (iv) and Lemma 2.1 implies that for $n \in \mathbb{N}$

$$E(V_{n+1} | \alpha_n) \leq (1 - 2\alpha r_n + \beta^2 r_n^2) ||\Delta_n||^2 + U_n.$$

The fact that $(r_n)_n$ is a null sequence according to (9) implies that there is an index n_o with

$$E(V_{n+1} | \alpha_n) \leq ||\Delta_n||^2 + U_n - V_n \tag{v}$$

for all $n \geq n_o$. Hence, $(V_{n+1}, \alpha_n)_{n \geq n_o}$, too, is a supermartingale.

Because of (iii), (iv) and (v) there are - according to the convergence theorem for supermartingales (cf. [1], Theorem 60.1) - random variables U_∞ und V_∞ such that with probability 1

$$\lim_{n \to \infty} U_n - U_\infty \quad \text{and} \quad \lim_{n \to \infty} V_n - V_\infty$$

and therefore

$$\lim_{n \to \infty} ||\Delta_n||^2 - V_\infty - U_\infty \quad \text{almost sure.} \tag{vi}$$

Because of a) $(||\Delta_n||^2)_n$ stochastically converges to 0, hence according to (vi), $(||\Delta_n||^2)_n$ converges to 0 almost sure.

The proposition a) of Theorem 3.1 can be formulated more precisely:

<u>Theorem 3.2</u> For k=1,...,κ let us assume

$$\limsup_{\mathbb{N}^{(k)} \ni n \to \infty} E||Z_n||^2 \le \sigma_k^2 \tag{10}$$

with constants $\sigma_1,...,\sigma_\kappa$. Hence

$$\limsup_{n \to \infty} n \cdot E||\Delta_n||^2 \le \frac{\sum\limits_{k=1}^{\kappa} q_k (M^{(k)} \sigma_k)^2}{2\alpha \cdot \bar{M} - 1}. \tag{11}$$

<u>Proof.</u> Assume an arbitrarily selected $\epsilon \in (0, 2\alpha - \frac{1}{\bar{M}})$. In analogy to

the proof of Theorem 3.1, Part a), for h: = $2 \cdot \alpha - \epsilon$ there is an

index n_1 with

$$E||\Delta_{n+1}||^2 \le (1 - h \cdot r_n) E||\Delta_n||^2 + r_n^2 E||Z_n||^2 \tag{i}$$

for all $n \ge n_1$. Because of (9) and (10) there is an index $n_0 \ge n_1$ with

$$M_m^2 E||Z_m||^2 \le (M^{(k)} \sigma_k)^2 + \epsilon \tag{ii}$$

for all $m \ge n_0$, $m \in \mathbb{N}^{(k)}$ for some $k \in \{1,...,\kappa\}$.

(i) and (ii) yield for all $n \ge n_0$

$$E||\Delta_{n+1}||^2 \le b_{n_0,n} E||\Delta_{n_0+1}||^2 + \sum_{m=n_0+1}^{n} b_{m,n} r_m^2 E||Z_m||^2$$

$$\le b_{n_0,n} E||\Delta_{n_0+1}||^2 + \sum_{k=1}^{\kappa} [(M^{(k)} \sigma_k)^2 + \epsilon] \cdot \sum_{m \in \mathbb{N}_{n_0,n}^{(k)}} b_{m,n} \frac{1}{m^2},$$

where $b_{m,n}: = \prod\limits_{j=m+1}^{n} (1 - \frac{hM_j}{j})$ and $\mathbb{N}_{n_0,n}^{(k)}: = \{n_0+1,...,n\} \cap \mathbb{N}^{(k)}$.

Hence, according to Lemma A.7 we obtain for A: $= h\bar{M} - 1$ and t:=1

$$\limsup_{n \to \infty} n \cdot E||\Delta_{n+1}||^2 \le \frac{\sum\limits_{k=1}^{\kappa} [(M^{(k)} \sigma_k)^2 + \epsilon] \cdot q_k}{(2\alpha - \epsilon) \cdot \bar{M} - 1}.$$

By the limiting process $\epsilon \to 0$ we obtain the estimate (11).

For given numbers $\alpha, q_1,...,q_\kappa$, $\sigma_1,...,\sigma_\kappa$ fulfilling condition (9.3),

the right hand side of inequality (11) is globally minimal, where

$$M^{(k)}: = \frac{1}{\sigma_k^2} \cdot [\alpha(\frac{q_1}{\sigma_1^2} + ... + \frac{q_\kappa}{\sigma_\kappa^2})]^{-1}, \quad k=1,...,\kappa. \tag{12}$$

Thus we obtain

<u>Corollary 3.1</u>. If the limits $M^{(1)},\ldots,M^{(\kappa)}$ given in (9.2) are chosen according to (12) and if condition (10) is fulfilled, then $\bar{M} = \alpha^{-1}$ and

$$\limsup_{n\to\infty} n\cdot E||\Delta_n||^2 \le [\alpha^2(\frac{q_1}{\sigma_1^2}+\ldots+\frac{q_\kappa}{\sigma_\kappa^2})]^{-1}. \tag{13}$$

Theorems 3.1 and 3.2 suggest that the sequence $(\sqrt{n}\cdot\Delta_n)_n$ under certain additional conditions converges towards a random vector. This conjecture is confirmed by the following theorem which generalizes the relevant propositions on the Robbins-Monro process described in [13] and [4].

<u>Theorem 3.3</u> Let $W^{(1)},\ldots,W^{(\kappa)}$ be symmetric matrices. For the sequence of estimation errors $(Z_n)_n$ we assume that

$$W^{(k)} = \lim_{\mathbb{N}^{(k)} \ni m\to\infty} E(Z_m Z_m^T|\mathcal{O}_m) \quad \text{a.s.} \tag{14.1}$$

for $k=1,\ldots,\kappa$,

$$\lim_{m\to\infty} E(||Z_m||^2 I_{(||Z_m||^2>t\,m)}|\mathcal{O}_m) = 0 \quad \text{a.s.} \tag{14.2}$$

for all $t>0$.

$(\sqrt{n}\cdot\Delta_n)_n$ will then converge in distribution to an $N(o,V)$-distributed random vector. The covariance matrix V will be calculated using the formula

$$V = P\,\hat{V}\,P^T, \tag{15.1}$$

where

$$\hat{V}: = \sum_{k=1}^{\kappa} q_k\,M^{(k)^2}\cdot(\frac{\hat{w}_{ij}^{(k)}}{\bar{M}(h_i+h_j)-1})_{1\le i,j\le\nu}, \tag{15.2}$$

$$\hat{W}^{(k)} = (\hat{w}_{ij}^{(k)})_{ij}: = P^T W^{(k)} P \quad \text{for } k=1,\ldots,\kappa. \tag{15.3}$$

Moreover, $h_1,\ldots,h_\nu$ are the eigenvalues of H, and P is the orthogonal matrix according to (6).

Proof. Let D be an arbitrary random vector having an $N(o,V)$-distribution. Because of Lemma 2.2, Lemma 2.3, Theorem 3.1, and with $r_n := \sqrt{n}$ it is sufficient to prove that

a) $\lim\limits_{n \to \infty} \sqrt{n} \ ||B_{m,n}|| = 0$ for all $m \in \mathbb{N}_o$,

b) there is a number L with

$$\sqrt{n} \sum_{m=1}^{n} ||B_{m,n}|| \ r_m \frac{1}{\sqrt{m}} \leq L \text{ for all } n \in \mathbb{N},$$

c) $(\sqrt{n} \sum_{m=1}^{n} B_{m,n} \ r_m \ Z_m)_n$ converges in distribution to D.

For $m \leq n$ set $c_{m,n} := \prod\limits_{j=m+1}^{n} (1 - r_j \cdot \alpha)$.

(7), (9) und (5) yield an $n_1 \in \mathbb{N}$ with

$$||B_{m,n}|| = \max_{1 \leq i \leq \nu} |B_{m,n}^{(i)}| \leq c_{m,n} \tag{i}$$

for all m,n where $n_1 \leq m \leq n$.

Thus, conditions a) and b) follow from (i), the relation $B_{m,n} = B_{m,n_1} \cdot B_{n_1,n}$ for $m < n_1 \leq n$, (9) und Lemma A7 with $t := \frac{1}{2}$, $A := \bar{M}\alpha - \frac{1}{2}$.

As to c), because of (14.1), Theorem B.1 and Theorem B.2 we only have zu show that the assumptions (B2) and (B4) listed in Annex B are satisfied; to this end we define

$$A_{m,n} := \sqrt{n} \ B_{m,n} \ r_m, \quad m \leq n.$$

(B2.1) follows from a). According to (i) we have for $n_1 \leq m \leq n$

$$||A_{m,n}|| \leq \hat{M} c_{m,n} \sqrt{\frac{n}{m}} \frac{1}{\sqrt{m}} =: a_{m,n}, \tag{ii}$$

where $\hat{M} := \sup\limits_{n} M_n$. Lemma A.7 with $t := 1$, $A := 2\bar{M}\alpha - 1$ yields

$$\lim_{n \to \infty} \sum_{m=n_1}^{n} a_{m,n}^2 < \infty.$$

Hence, (B 2.2) und (B 2.3) are satisfied.

Because of Lemma A.7 with $t := -\frac{1}{2}$ and $A := \bar{M}\alpha - \frac{1}{2}$ and because of Lemma A.5a) there is - for small values of $\delta > 0$ - an index $n_o \geq n_1$ with

$$c_{m,n} \sqrt{\frac{n}{m}} \leq 2 \cdot \left(\frac{m}{n}\right)^{\bar{M}\alpha - \frac{1}{2} - \delta} \leq 2$$

for all $n_o \leq m \leq n$. Hence, because of (ii) we have

$$||A_{m,n}|| \leq \frac{2\hat{M}}{\sqrt{m}} \text{ for } n_o \leq m \leq n.$$

Thus, (B 2.4) with $\alpha_m := \frac{2\hat{M}}{\sqrt{m}}$ is also satisfied, and (B 4) follows from (14.2).

It remains to be shown that (15) and (B 2.5) refer to identical matrices V. This is true if for each $k \in \{1,\ldots,\kappa\}$

$$\lim_{n \to \infty} n \cdot \sum_{m \in \mathbb{N}_{o,n}^{(k)}} \hat{B}_{m,n} \hat{W}^{(k)} \hat{B}_{m,n} r_m^2 = \tag{iii}$$

$$= q_k M^{(k)2} \cdot \left(\frac{\hat{w}_{ij}^{(k)}}{\check{M}(h_i+h_j)-1} \right)_{1 \leq i,j \leq \nu},$$

where $\mathbb{N}_{o,n}^{(k)} := \{1,\ldots,n\} \cap \mathbb{N}^{(k)}$.

Now we start from chosen values $k \in \{1,\ldots,\kappa\}$ and $i,j \in \{1,\ldots,\nu\}$.

(5) and (9.3) imply $A := \check{M} \cdot (h_i+h_j)-1 > 0$; hence, according to (8),(9), Lemma A.2, and Lemma A.7 we have

$$\lim_{n \to \infty} \sum_{m \in \mathbb{N}_{o,n}^{(k)}} B_{m,n}^{(i)} \cdot B_{m,n}^{(j)} \cdot \left(\frac{n}{m}\right)^1 \cdot \frac{M_m^2}{m} = \frac{q_k M^{(k)2}}{\check{M}(h_i+h_j)-1} .$$

Hence, (iii) is true with respect to the (i,j)-th components of the matrices.

The above proof and Theorem B.2 confirm the validity of the proposition expressed in Theorem 3.3 in the case that - instead of (14) - the following assumptions are satisfied for all $k=1,\ldots,\kappa$ and $t>0$:

$$W^{(k)} = \lim_{\mathbb{N}^{(k)} \ni m \to \infty} E(Z_m Z_m^T), \tag{14.1)'}$$

$$\lim_{m \to \infty} E||E(Z_m Z_m^T | \mathcal{O}_m) - E Z_m Z_m^T|| = 0, \tag{14.2)'}$$

$$\lim_{m \to \infty} E(||Z_m||^2 I_{(||Z_m||^2 > t \cdot m)}) = 0 \tag{14.3)'}$$

If $Z_1, Z_2, \ldots$ are independent and distributed like a given random vector Z with a finite covariance matrix $W := E Z Z^T$, assumptions (14)' are satisfied by $W^{(k)} = W$ for $k=1,\ldots,\kappa$. An example of a sequence of estimation errors $(Z_n)_n$ which satisfies assumptions (14) is given in section 6.

By carrying out a simple method of calculation, we find that the covariance matrix V given by (15) is the solution of the matrix equation

$$(\tilde{M}H - \frac{1}{2}I) \cdot V + V \cdot (\tilde{M}H - \frac{1}{2}I) = \sum_{k=1}^{\kappa} q_k\, M^{(k)2}\, W^{(k)}. \qquad (15)'$$

Because of (5) and (9.3) the Matrix $\tilde{M}H - \frac{1}{2} \cdot I$ is positive definite, and thus, (15)' has one and only one solution.

For the "mean Z-covariance matrix"

$$\tilde{W}: = \sum_{k=1}^{\kappa} q_k \cdot (\frac{M^{(k)}}{\tilde{M}})^2 \cdot W^{(k)},$$

(15)' may be written as follows

$$(\tilde{M} \cdot H - \frac{1}{2}I) \cdot V + V \cdot (\tilde{M}H - \frac{1}{2}I) = \tilde{M}^2 \cdot \tilde{W}.$$

Assuming the decomposition $\mathbb{N} = \mathbb{N}^{(1)}$, according to (8) we obtain the

Corollary 3.2 (J. Sacks)

Let W be a symmetric matrix and M>o.

If

a) $W = \lim_{n \to \infty} E(Z_n Z_n^T | \mathcal{O}_n)$ almost everywhere or $W = \lim_{n \to \infty} E(Z_n Z_n^T)$,

b) $(Z_n)_n$ satisfies (14.2) or the two assumptions (14.2)' and (14.3)',

c) $r_n = \frac{M_n}{n}$ for $n=1,2,\ldots,$

d) $\lim_{n \to \infty} M_n = M > \frac{1}{2\alpha}$,

then $(\sqrt{n} \cdot \Delta_n)_n$ converges in distribution to an N(o,V)-distributed random vector. The covariance matrix V is given by

$$(MH - \frac{1}{2}I) \cdot V + V(MH - \frac{1}{2}I) = M^2 \cdot W.$$

Proof. Choose $\kappa: = 1$ in (8.1) and take $M^{(1)}: = M$, $W^{(1)}: = W$. The assertion follows from Theorem 3.3 and the corresponding remarks.

4. **Semi-stochastic approximation procedures**

We now start from a given disjoint decomposition

$$\mathbb{N} - \mathbb{N}^{(1)} \cup \mathbb{N}^{(2)} \tag{16}$$

of $\mathbb{N}$ assuming

$$Z_n \equiv 0 \text{ for all } n \in \mathbb{N}^{(1)}, \tag{17}$$

i.e. in the so-called <u>deterministic step</u> $n \in \mathbb{N}^{(1)}$ no error of estimation is made when calculating $G(X_n)$.

In terms of the step size r_n this section starts from the following approach:

$$r_n - \begin{cases} r & , & \text{für } n \in \mathbb{N}^{(1)} \\ M_n \cdot R_n & , & \text{für } n \in \mathbb{N}^{(2)} \end{cases}, \tag{18.1}$$

where

$$r \in (0, \frac{2\alpha}{\beta^2}), \tag{18.2}$$

$(M_n)_n$ is a positive sequence with

$$\hat{M} : - \sup_{n \in \mathbb{N}^{(2)}} M_n < \infty, \tag{18.3}$$

$(R_n)_n$ is a positive null sequence. \hfill (18.4)

Because of (18.2) we have

$$p : - 1 - 2\alpha r + \beta^2 \cdot r^2 \in [0,1), \tag{19}$$

which plays an important role in further convergence analyses. An initial proposition on the sequence $(E||\Delta_n||^2)_n$ is given by

<u>Lemma 4.1.</u> There is an index n_0 such that

$$E||\Delta_{n+1}||^2 \le b_{n_0,n} \cdot E||\Delta_{n_0+1}||^2 + (\hat{M}\sigma)^2 \cdot \sum_{m \in \mathbb{N}^{(2)}_{n_0,n}} b_{m,n} \cdot R_m^2$$

for all $n \ge n_0$ assuming

$$b_{m,n} : - p^{|\mathbb{N}^{(1)}_{m,n}|} \text{ for } m \le n \text{ with } p \text{ as defined by (19).}$$

<u>Proof.</u> Lemma 2.1, (4.2), (17) and (18) yield

$$E||\Delta_{n+1}||^2 \le (1-2\alpha r_n + \beta^2 r_n^2)E||\Delta_n||^2 + \begin{cases} 0 & \text{, for } n \in \mathbb{N}^{(1)} \\ r_n^2(\sigma^2 + \gamma^2 E||\Delta_n||^2), & \text{for } n \in \mathbb{N}^{(2)} \end{cases}$$

$$\le \begin{cases} p \cdot E||\Delta_n||^2 & \text{, for } n \in \mathbb{N}^{(1)} \\ (1-2\alpha r_n + (\beta^2+\gamma^2)r_n^2)E||\Delta_n||^2 + (\hat{M}\sigma)^2 R_n^2 & \text{, for } n \in \mathbb{N}^{(2)} \end{cases}$$

for all $n \in \mathbb{N}$. Because of (18) $(r_n)_{n \in \mathbb{N}^{(2)}}$ is a null sequence, and hence, there is an index n_o with

$$E||\Delta_{n+1}||^2 \le \begin{cases} p \cdot E||\Delta_n||^2 & \text{, for } n \in \mathbb{N}^{(1)} \\ E||\Delta_n||^2 + (\hat{M}\sigma)^2 \cdot R_n^2, & \text{otherwise} \end{cases}$$

for all $n \ge n_o$. The assertion of this lemma follows by complete induction.

According to Lemma 4.1 the convergence behavior of $(E||\Delta_n||^2)_n$ is mainly determined by the sequence $(R_n)_n$.

4.1 Optimal step size

As shown in [10], by optimal step size selection we obtain

$$E||\Delta_n||^2 - O((1 - \frac{\alpha^2}{\beta^2})^{|\{1,\ldots,n\} \cap \mathbb{N}^{(1)}|}).$$

As to the optimal step size $(r_n^*)_n$ the following holds true:

$$r_n^* - \begin{cases} \frac{\alpha}{\beta^2} & \text{, for } n \in \mathbb{N}^{(1)} \\ O(E||\Delta_n||^2) & \text{, for } n \in \mathbb{N}^{(2)} \end{cases}.$$

Hence, $(r_n^*)_n$ takes the form defined by (18) with

$$r - \frac{\alpha}{\beta^2} \text{ and } R_n - (1 - \frac{\alpha^2}{\beta^2})^{|\{1,\ldots,n\} \cap \mathbb{N}^{(1)}|}.$$

Let us assume the sequence $(R_n)_n$ in (18) to be given in the more general form

$$R_n: - p^{|\{1,\ldots,n\} \cap \mathbb{N}^{(1)}|}, \quad n-1,2,\ldots \tag{20}$$

with p as defined by (19).

This step size, too, yields an adequate estimation for $E||\Delta_n||^2$:

<u>Theorem 4.1.</u> Let $\mathbb{N}^{(1)}$ be such that

$$\sum_{n \in \mathbb{N}} R_n < \infty. \tag{21}$$

Then we obtain

a) $E||\Delta_n||^2 = O(R_n)$

b) $\lim_{n \to \infty} ||\Delta_n|| = 0$ almost sure.

<u>Proof.</u> According to Lemma 4.1 there is an $n_o \in \mathbb{N}$ with

$$\frac{E||\Delta_{n+1}||^2}{R_n} \leq \frac{b_{n_o,n}}{R_n} E||\Delta_{n_o+1}||^2 + (\hat{M}\sigma)^2 \cdot \sum_{m \in \mathbb{N}^{(2)}_{n_o,n}} \frac{b_{m,n} R_m}{R_n} R_m$$

for all $n \geq n_o$, where $b_{m,n} := p^{|\mathbb{N}^{(1)}_{m,n}|}$.

Since $b_{m,n} \cdot R_m = R_n$ for $m \leq n$, we have

$$\frac{E||\Delta_{n+1}||^2}{R_n} \leq \frac{E||\Delta_{n_o+1}||^2}{R_{n_o}} + (\hat{M}\sigma)^2 \cdot \sum_{m \in \mathbb{N}} R_m,$$

for $n \geq n_o$; according to (21) we obtain

$$E||\Delta_n||^2 \leq L R_n \quad \text{for all } n \in \mathbb{N} \tag{i}$$

with a positive number L. This result corresponds to assertion a).

Because of Chebyshev's inequality and according to (i) we have

$$\mathbb{P}(\bigcup_{n \geq k} (||\Delta_n|| > \epsilon)) \leq \sum_{n \geq k} \mathbb{P}(||\Delta_n|| > \epsilon) \leq \frac{L}{\epsilon^2} \sum_{n \geq k} R_n.$$

for each $\epsilon > 0$ and $k \in \mathbb{N}$.

The assertion b) follows from this and from (21).

Assumption (21) is definitely satisfied if

$$q := \liminf_{n \to \infty} \frac{|\{1, \ldots, n\} \cap \mathbb{N}^{(1)}|}{n} > 0, \tag{22}$$

because this case yields an index n_o with

$$R_n = (p^{\frac{|\{1, \ldots, n\} \cap \mathbb{N}^{(1)}|}{n}})^n \leq (p^{\frac{q}{2}})^n$$

for all $n \geq n_o$.

The following theorem provides a proposition on the convergence behavior of the sequence ($\frac{1}{\sqrt{R_n}} \Delta_n)_n$. In this context we refer to

Lemma 4.2. There is an index n_o such that for all m,n with $n_o \leq m \leq n$

$$||B_{m,n}|| \leq \frac{\sqrt{R_n}}{\sqrt{R_m}} \hat{p}^{|\mathbb{N}_{m,n}^{(1)}|}$$

where $\hat{p}: = \frac{\max(|1-r\alpha|,|1-r\cdot\beta|)}{\sqrt{p}} \in [0,1]$.

$\hat{p}<1$ for $\alpha<\beta$.

Proof. Because of (5) and (19) we have

$$(1-rh_i)^2 \leq \max((1-r\alpha)^2,(1-r\beta)^2) \underset{(-)}{<} p$$

for all eigenvalues $h_1,\ldots,h_\nu$ of H.

If $n_o \in \mathbb{N}$, then

$$r_n \cdot h_i \leq 1 \text{ for all } i=1,\ldots,\nu, \ n_o \leq n \in \mathbb{N}^{(2)}.$$

Hence, for all $n_o \leq m \leq n$

$$||B_{m,n}||^2 - \max_{1\leq i\leq\nu} B_{m,n}^{(i)2} \leq \max_{1\leq i\leq\nu} (1-rh_i)^{2|\mathbb{N}_{m,n}^{(1)}|}$$

$$\leq \frac{[\max(|1-r\alpha|,|1-r\beta|)]^{2|\mathbb{N}_{m,n}^{(1)}|}}{p^{|\mathbb{N}_{m,n}^{(1)}|}} \cdot \frac{p^{|\mathbb{N}_{o,n}^{(1)}|}}{p^{|\mathbb{N}_{o,m}^{(1)}|}} -$$

$$- \hat{p}^{2|\mathbb{N}_{m,n}^{(1)}|} \cdot \frac{R_n}{R_m}.$$

Therem 4.2. If $\alpha<\beta$ and there are numbers $q_1>0$ und $q_o \in \mathbb{R}$ with

$$|(m+1,\ldots,n)\cap\mathbb{N}^{(1)}| \geq q_1\cdot(n-m) - q_o$$

for all $m\leq n$, then the sequence ($\frac{1}{\sqrt{R_n}} \Delta_n)_n$ converges stochastically to 0.

Proof. According to Lemma 2.2, Lemma 2.3, and Theorem 4.1 we only have to show that

a) $\lim\limits_{n \to \infty} \dfrac{1}{\sqrt{R_n}} \, ||B_{m,n}|| = 0$ for all $m \in \mathbb{N}_o$,

b) there is a number L with

$$\frac{1}{\sqrt{R_n}} \sum_{m=1}^{n} ||B_{m,n}|| \; r_m \sqrt{R_m} \le L \quad \text{for all } n \in \mathbb{N},$$

c) $\lim\limits_{n \to \infty} \dfrac{1}{\sqrt{R_n}} \sum\limits_{m=1}^{n} ||B_{m,n}|| \, r_m \, E||Z_m|| = 0.$

Let n_o and $\hat{p} \in [0,1)$ be defined as in Lemma 4.2. Due to the assumption of the theorem and Lemma 4.2 we have

$$||B_{m,n}|| \le \frac{\sqrt{R_n}}{\sqrt{R_m}} \cdot (\hat{p}^{q_1})^{n-m} \cdot \hat{p}^{-q_o} \tag{i}$$

for $n_o \le m \le n$.

For $m < n_o \le n$ we have

$$B_{m,n} = B_{m,n_o} \cdot B_{n_o,n} \; ,$$

yielding assertion a) according to (i).

For $n > n_o$ (i) yields

$$\frac{1}{\sqrt{R_n}} \sum_{m=n_o+1}^{n} ||B_{m,n}|| \, r_m \sqrt{R_m} \le (\sup_k r_k) \cdot \hat{p}^{-q_o} \cdot \sum_{m=n_o+1}^{n} (\hat{p}^{q_1})^{n-m} \; ,$$

resulting in condition b) according to a).

For $n > n_o$, (17),(18),(4.2) and (i) yield

$$\frac{1}{\sqrt{R_n}} \sum_{m=n_o+1}^{n} ||B_{m,n}|| \, r_m E||Z_m|| \le$$

$$\frac{1}{\sqrt{R_n}} \sum_{m \in \mathbb{N}_{n_o,n}^{(2)}} ||B_{m,n}|| \, M_m \, R_m \sqrt{\sigma^2 + \gamma^2 \sup_k E||\Delta_k||^2} \le$$

$$\hat{M} \sqrt{\sigma^2 + \gamma^2 \sup_k E||\Delta_k||^2} \; \hat{p}^{-q_o} \sum_{m \in \mathbb{N}_{n_o,n}^{(2)}} (\hat{p}^{q_1})^{n-m} \sqrt{R_m} \; .$$

Since $\lim\limits_{m \to \infty} \sqrt{R_m} = 0$, Lemma A.2.c3 yields

$$\lim_{n \to \infty} \frac{1}{\sqrt{R_n}} \sum_{m=n_o+1}^{n} ||B_{m,n}|| \, r_m E||Z_m|| = 0,$$

from which condition c) follows once again according to a).

4.2 Non-optimal step size

This section differs from the previous subsection in that the sequence $(R_n)_n$ is assumed to be given by the following equation instead of (20):

$$R_n := \frac{1}{n} \ , \ n=1,2,\ldots \tag{23}$$

Furthermore, we start from the set $\mathbb{N}^{(1)}$ such that for all $m \le n$

$$|(m+1,\ldots,n) \cap \mathbb{N}^{(1)}| \ge q_1(n-m) - q_0 \tag{24}$$

where $q_1 > 0$ and $q_0 \in \mathbb{R}$ are constants.

In analogy to Theorem 3.1 and 4.1 we now have

Theorem 4.3. The following holds true

a) $E||\Delta_n||^2 = 0(\frac{1}{n^2})$

b) $\lim\limits_{n \to \infty} ||\Delta_n|| = 0$ almost sure.

Proof. Let n_0 be as defined in Lemma 4.1. Then, Lemma 4.1, (23) and (24) yield for all $n \ge n_0$

$$n^2 \cdot E||\Delta_{n+1}||^2 \le p^{-q_0}[n^2(p^{q_1})^{n-n_0} E||\Delta_{n_0+1}||^2 +$$

$$+ (\hat{M}\sigma)^2 \cdot \sum_{m \in \mathbb{N}_{n_0,n}^{(2)}} (p^{q_1})^{n-m}(\frac{n}{m})^2 \ ].$$

Hence, according to Lemma A.8 we have

$$\limsup_{n \to \infty} n^2 \ E||\Delta_n||^2 \le p^{-q_0} \frac{(\hat{M}\sigma)^2}{1-p^{q_1}} < \infty.$$

Condition a) results in $\sum\limits_{n=1}^{\infty} E||\Delta_n||^2 < \infty$, from which the assertion given in b) follows according to Chebyshev's inequality.

The proposition on the convergence behavior for the case of $\mathbb{N}^{(1)} = (1, 3, 5,\ldots)$ as given in Theorem 4.3a) has already been verified in [9]. Theorem 4.5 shows that the sequence $(n \cdot \Delta_n)_n$ is not generally characterized by convergence in variation. We arrive, however, at

<u>Corollary 4.1.</u> For $\mu < 1$, $(n^{\mu} \cdot \Delta_n)_n$ stochastically converges to 0.

<u>Proof.</u> The assertion follows from Theorem 4.3a) and Chebyshev's inequality.

The following theorem provides information on the limiting distribution of $(n \cdot \Delta_n)_n$:

<u>Theorem 4.4.</u> Für each $k \in \mathbb{N}_0$

$$(n \cdot \Delta_{n+1} + \sum_{m \in \mathbb{N}_{k,n}^{(2)}} B_{m,n} \cdot \frac{n}{m} M_m Z_m)_{n > k}$$

stochastically converges to 0.

<u>Proof.</u> Because of Lemma 2.2., Lemma 2.3, Theorem 4.3, and (17) we have to show that

a) $\lim_{n \to \infty} n \cdot ||B_{m,n}|| = 0$ for all $m \in \mathbb{N}_0$

b) there is a number L with

$$n \cdot \sum_{m=1}^{n} ||B_{m,n}|| \cdot r_m \cdot \frac{1}{m} \leq L \text{ for all } n \in \mathbb{N}.$$

In analogy to the proof of Lemma 4.2 we will demonstrate the existence of an index n_0 with

$$||B_{m,n}|| \leq (\sqrt{p})^{|\mathbb{N}_{m,n}^{(1)}|} \leq (\sqrt{p})^{q_1(n-m)-q_0} \tag{i}$$

for all $n_0 \leq m \leq n$ from which condition a) follows.

Furthermore (i) yields for $n \geq n_0$

$$n \cdot \sum_{m=n_0+1}^{n} ||B_{m,n}|| \, r_m \frac{1}{m} \leq$$

$$(\sup_k r_k) \cdot p^{-\frac{q_0}{2}} \sum_{m=n_0+1}^{n} (p^{\frac{q_1}{2}})^{n-m} (\frac{n}{m}) \, .$$

From this condition b) follows according to Lemma A.8 and a).

Propositions on the limiting distribution of $(n \cdot \Delta_n)_n$ may be concretized if sets $\mathbb{N}^{(1)}$ and $\mathbb{N}^{(2)}$ are of the following simple category:

$$\mathbb{N}^{(2)} = \{\bar{n}+k\cdot(n_S+n_D) + \ell : k \in \mathbb{N}_o, \ \ell \in (1,\ldots,n_S)\}$$

(25)

$$\mathbb{N}^{(1)} = \mathbb{N} \setminus \mathbb{N}^{(2)}$$

with constant $\bar{n}, n_S, n_D \in \mathbb{N}$ i.e. n_S stochastic steps from $\mathbb{N}^{(2)}$ and n_D deterministic steps from $\mathbb{N}^{(1)}$ are taken alternately. For $t \in (1,\ldots,n_S+n_D)$ we consider the set

$$\mathbb{M}_t := \{\bar{n}+k\cdot(n_S+n_D)+t : k \in \mathbb{N}_o\}.$$

$\mathbb{M}_t$ is the set of all <u>t-th stochastic steps</u> for $t \leq n_S$ and the set of all <u>$(t-n_S)$-th deterministic steps</u> for $t > n_S$. The next theorem yields the formula for the limit covariance matrix of $((n+1)\cdot\Delta_{n+1})_n$ with the limit being formed on the basis of $n \in \mathbb{M}_t$. For this purpose we refer to

<u>Lemma 4.3.</u> Let $y \in \mathbb{R}$ with $|y| < 1$ and $(A_n)_{n \in \mathbb{N}^{(2)}}$ be a sequence with

$$\underline{A} := \liminf_{\mathbb{N}^{(2)} \ni n \to \infty} A_n > 0, \quad \bar{A} := \limsup_{\mathbb{N}^{(2)} \ni n \to \infty} A_n < \infty.$$

For $M,N \in \mathbb{N}$ where $M \leq N$ take

$$b_{M,N} := y^{|\mathbb{N}_{M,N}^{(1)}|} \cdot \prod_{j \in \mathbb{N}_{M,N}^{(2)}} (1 - \frac{A_j}{j}).$$

Under assumption (25) we obtain

a) $\lim_{N \to \infty} N^c \, b_{K,N} = 0$

b) $\lim_{\mathbb{M}_t \ni N \to \infty} \sum_{M \in \mathbb{N}_{K,N}^{(2)}} b_{M,N} \cdot (\frac{N}{M})^c =$

$$= y^{\max(0,t-n_S)} \cdot [\frac{n_S}{1-y^{n_D}} - \max(0,n_S-t)].$$

for all $c \in \mathbb{R}$, $K \in \mathbb{N}_o$ and $t \in (1,\ldots,n_S+n_D)$.

<u>Proof.</u> Due to (25) we have

$$|\mathbb{N}_{K,N}^{(1)}| \geq \frac{n_D}{n_S+n_D}\cdot(N-K) - (n_S+n_D)$$

for $K \leq N$. Since $|y| < 1$, we thus obtain assertion a). Furthermore,

$$q_2: = \lim_{j \to \infty} \frac{|\mathbb{N}_{o,j}^{(2)}|}{j} - \frac{n_S}{n_S + n_D} > 0.$$

Hence, under the assumption concerning $(A_n)_n$ there is an index N_o with

$$\prod_{j \in \mathbb{N}_{M,N}^{(2)}} (1 - \frac{2\,\bar{A}\,q_2}{|\mathbb{N}_{o,j}^{(2)}|}) \leq \prod_{j \in \mathbb{N}_{M,N}^{(2)}} (1 - \frac{A_j}{j}) \leq$$

$$\leq \prod_{j \in \mathbb{N}_{M,N}^{(2)}} (1 - \frac{\frac{1}{2}\,\underline{A}\,q_2}{|\mathbb{N}_{o,j}^{(2)}|}) \quad \text{for all } N_o \leq M \leq N. \tag{i}$$

For positive values of A, Lemma A.5a) and Lemma A.6a) yield

$$\prod_{j \in \mathbb{N}_{M,N}^{(2)}} (1 - \frac{A}{|\mathbb{N}_{o,j}^{(2)}|}) - \prod_{i=|\mathbb{N}_{o,M}^{(2)}|+1}^{|\mathbb{N}_{o,N}^{(2)}|} (1 - \frac{A}{i}) \sim$$

$$\sim (\frac{|\mathbb{N}_{o,M}^{(2)}|}{|\mathbb{N}_{o,N}^{(2)}|})^A \sim (\frac{M}{N})^A,$$

and according to (i) we obtain

$$(\frac{M}{N})^a \ll \prod_{j \in \mathbb{N}_{M,N}^{(2)}} (1 - \frac{A_j}{j}) \cdot (\frac{N}{M})^c \ll (\frac{M}{N})^b \tag{ii}$$

where $a: = 2 \cdot \bar{A} \cdot q_2 - c$, $b: = \frac{1}{2} \cdot \underline{A} \cdot q_2 - c$.

Assertion a) enables us to assume $K \in \mathbb{N}^{(1)}$ and $K+1 \in \mathbb{N}^{(2)}$.

$N \geq K$, $N \in \mathbb{M}_t$ and $M \in \mathbb{N}_{K,N}^{(2)}$ may then be written as follows:

$$N - K + n(n_S + n_D) + t =: N(n)$$

$$M - K + m(n_S + n_D) + i =: M(m,i)$$

where $n \in \mathbb{N}_o$, $m \in \{0, \ldots, n\}$, $i \in \{1, \ldots, n_S\}$ and $i \leq t$ for $m - n$.

Hence

$$|\mathbb{N}_{M,N}^{(1)}| - (n-m) \cdot n_D + t_D$$

where $t_D: = \max(0, t - n_S)$.

For $m \leq n$ we take

$$a_{m,n}: - \frac{1}{\bar{n}(m)} \cdot \sum_{i=1}^{\bar{n}(m)} \prod_{j \in \mathbb{N}_{M(m,i),N(n)}^{(2)}} (1 - \frac{A_j}{j}) \cdot (\frac{N(n)}{M(m,i)})^c,$$

where $\bar{n}(m): - \begin{cases} n_S, & \text{for } m<n \\ t_S, & \text{for } m=n \end{cases}$

and $t_S: - \min(n_S, t)$. Hence

$$\sum_{M \in \mathbb{N}_{K,N(n)}^{(2)}} b_{M,N} \left(\frac{N}{M}\right)^c - \sum_{m=o}^{n} y^{(n-m)n_D+t_D} \cdot \bar{n}(m) \cdot a_{m,n}$$

(iii)

$$- y^{t_D} \cdot [n_S \cdot \sum_{m=o}^{n} (y^D)^{n_D \, n-m} a_{m,n} - (n_S - t_S) \cdot a_{n,n}]$$

Because of (ii) we get

$$\left(\frac{m}{n}\right)^a \ll a_{m,n} \ll \left(\frac{m}{n}\right)^b,$$

and hence, according to Lemma A.8 and (iii) we have

$$\lim_{\mathbb{M}_t \ni N \to \infty} \sum_{M \in \mathbb{N}_{K,N}^{(2)}} b_{M,N} \cdot \left(\frac{N}{M}\right)^c - y^{t_D} \cdot [\frac{n_S}{1-y^{n_D}} - (n_S - t_S)] .$$

Theorem 4.5. If assumption (25) is satisfied and there are given

limits

$$M_{Stoch}: - \lim_{\mathbb{N}^{(2)} \ni n \to \infty} M_n > 0 \qquad (26)$$

$$W^{(2)} : - \lim_{\mathbb{N}^{(2)} \ni n \to \infty} E \, Z_n \, Z_n^T, \qquad (27)$$

then

$$\lim_{\mathbb{M}_t \ni n \to \infty} E \, ((n+1)^2 \cdot \Delta_{n+1} \, \Delta_{n+1}^T) - V_t ,$$

for every $t \in \{1, \ldots, n_S + n_D\}$ with matrix V_t being calculated using the

formula

$$V_t - P \, \hat{V}_t \, P^T, \qquad (28.1)$$

where

$$\hat{V}_t: - M_{Stoch}^2 \left(\hat{w}_{ij}^{(2)} \cdot P_{ij}^{\max(0, t-n_S)} [\frac{n_S}{1-P_{ij}} - \max(0, n_S - t)] \right)_{1 \le i, j \le \nu} \qquad (28.2)$$

$$\hat{W}^{(2)} - (\hat{w}_{ij}^{(2)})_{i,j}: - P^T \, W^{(2)} \, P \qquad (28.3)$$

$$P_{ij}: - (1-rh_i)(1-rh_j) \text{ for } 1 \le i, j \le \nu. \qquad (28.4)$$

$h_1, \ldots, h_\nu$ are again the eigenvalues of H and P is the orthogonal

matrix according to (6).

<u>Proof.</u> Due to Theorem 4.4, (4.1), und (7) we have to show that

$$\lim_{M_t \ni n \to \infty} \sum_{m \in \mathbb{N}_{o,n}^{(2)}} \hat{B}_{m,n} \cdot (P^T E(Z_m Z_m^T) P) \cdot \hat{B}_{m,n} \cdot (\frac{n}{m})^2 \cdot M_m^2 = \hat{V}_t,$$

which for the components means that

$$\lim_{M_t \ni n \to \infty} \sum_{m \in \mathbb{N}_{o,n}^{(2)}} B_{m,n}^{(i)} \cdot B_{m,n}^{(j)} \cdot (\frac{n}{m})^2 \cdot \hat{w}_{ij}(m) \cdot M_m^2 = \hat{v}_{ij} \qquad (i)$$

for all $1 \le i, j \le n$, where $\hat{V}_t = (\hat{v}_{ij})_{ij}$, and for $m \in \mathbb{N}^{(2)}$ we have

$$\hat{W}(m) = (\hat{w}_{ij}(m))_{ij} := P^T \cdot E(Z_m Z_m^T) \cdot P.$$

Choose fixed values of $i, j \in \{1, \ldots, v\}$. According to (18.1) and (23) we have

$$B_{m,n}^{(i)} \cdot B_{m,n}^{(j)} = P_{i,j}^{|\mathbb{N}_{m,n}^{(1)}|} \cdot \prod_{k \in \mathbb{N}_{m,n}^{(2)}} (1 - \frac{M_k h_i}{k}) \cdot (1 - \frac{M_k h_j}{k})$$

and $|p_{ij}| \le p < 1$ according to (5) and (19).

(26) und (27) yield

$$\lim_{\mathbb{N}^{(2)} \ni m \to \infty} \hat{w}_{ij}(m) \cdot M_m^2 = \hat{w}_{ij}^{(2)} \cdot M_{Stoch}^2.$$

Hence, due to Lemma 4.3 and Lemma A.2 we have

$$\lim_{M_t \ni n \to \infty} \sum_{m \in \mathbb{N}_{o,n}^{(2)}} B_{m,n}^{(i)} \cdot B_{m,n}^{(j)} \cdot (\frac{n}{m})^2 \cdot \hat{w}_{ij}(m) \cdot M_m^2 =$$

$$\hat{w}_{ij}^{(2)} \cdot M_{Stoch}^2 \cdot p_{ij}^{\max\{0, t-n_S\}} \cdot [\frac{n_S}{1-p_{ij}} - \max\{0, n_S - t\}],$$

i.e. (i) is satisfied in accordance with (28).

For $t = n_S$ the elements of matrix $\hat{V}_t$ in (28) assume their greatest absolute value. In this sense, with

$$V_{n_S} = n_S \cdot M_{Stoch}^2 \cdot p \, (\frac{\hat{w}_{ij}^{(2)}}{1-p_{ij}})_{i,j} \, p^T$$

we obtain the "greatest" matrix among $V_1, \ldots, V_{n_S + n_D}$.

V_{n_S} can be easily verified as the solution of the following matrix equation:

$$V_{n_S} - (I - rH)^{n_D} \cdot V_{n_S} \cdot (I - rH)^{n_D} = n_S \, M_{Stoch}^2 \, W^{(2)} \qquad (29)$$

For every small values of step size r of deterministic steps, i.e. for $r \cdot \beta \ll 1$, we obtain on the basis of (5)

$$(I - rH)^{n_D} \approx I - n_D r \, H,$$

and hence - according to (29) - we find

$$H \cdot V_{n_S} + V_{n_S} \cdot H \approx \frac{n_S \, M_{Stoch}^2}{n_D \cdot r} \cdot W^{(2)}.$$

5. Constant step size

This section starts from constant values of step size r_n in (2) i.e.

$$r_n \equiv r \quad \text{für } n=1,2,3\ldots \tag{30.1}$$

Since $(r_n)_n$ will no longer be a null sequence, sequence $(X_n)_n$ cannot be assumed to converge - in distribution - to x^*. In the square mean X_n will, however, lie - for great values of n - in a spherical neighborhood of x^* with the radius being proportional to $\sqrt{r}$.

For reasons of convergence we assume

$$r < \frac{2\alpha}{\beta^2 + \gamma^2} . \tag{30.2}$$

Theorem 5.1.

$$\limsup_{n \to \infty} E||\Delta_n||^2 \leq r \cdot \frac{\sigma^2}{2\alpha - (\beta^2 + \gamma^2)r}$$

Proof. Lemma 2.1 and (4.2) yield for every n

$$E||\Delta_{n+1}||^2 \leq (1-a) \cdot E||\Delta_n||^2 + a \cdot \frac{r^2 \sigma^2}{a}$$

where a: $= 2\alpha r - (\beta^2 + \gamma^2) \cdot r^2$.

According to Lemma A.4 we obtain the desired estimation.

According to Kushner and Huang (cf. [11]) the random vector $\frac{1}{\sqrt{r}} \Delta_n$

approximates an N(o,V)-distribution for very small values of r and

great values of n, where the covariance matrix V is given by the
equation

$$H \cdot V + V \cdot H = W.$$

W is the covariance matrix of the estimation error $Z(\omega)$ at x^*.

If $G(x)$ is affine linear, i.e. if (3) reads

$$\delta(x) \equiv 0 \text{ for all } x \in \mathbb{R}^{\nu}, \tag{31}$$

we arrive at a similar statement on the covariance matrices of the
limits of $(\frac{1}{\sqrt{r}} \Delta_n)_n$.

As in section 4 we start from a given decomposition of $\mathbb{N}$, i.e.

$$\mathbb{N} = \mathbb{N}^{(1)} \cup \mathbb{N}^{(2)} \tag{32}$$

Moreover, we assume given symmetric matrices $W^{(k)}$ and numbers
$u_k, v_k \geq 0$ for $k \in \{1, 2\}$ such that

$$E \, Z_n \, Z_n^T \leq W^{(k)} + (u_k \, E||\Delta_n|| + v_k \, E||\Delta_n||^2) \cdot I \tag{33}$$

for every $n \in \mathbb{N}^{(k)}$.

In the case of two symmetric matrices A,B the relation A≤B means that
B-A is positive semidefinite.

The following lemma is important in the present section:

<u>Lemma 5.1.</u> Fo every $n \in \mathbb{N}^{(k)}$ where $k \in \{1, 2\}$ we find

$$E \, \Delta_{n+1} \, \Delta_{n+1}^T \leq (I - rH) \cdot E \, \Delta_n \, \Delta_n^T \cdot (I - rH) +$$

$$+ r^2 \cdot [W^{(k)} + (u_k \, E||\Delta_n|| + v_k \, E||\Delta_n||^2) \cdot I] \tag{34}$$

<u>Proof.</u> Let $k \in \{1, 2\}$ and $n \in \mathbb{N}^{(k)}$. According to (2),(3) und (31) we have
$\Delta_{n+1} = (I - rH)\Delta_n - r \cdot Z_n$, hence - according to (4.1) -

$$E(\Delta_{n+1} \, \Delta_{n+1}^T | \mathcal{O}_n) = (I - rH)\Delta_n \, \Delta_n^T \cdot (I - rH) + r^2 E(Z_n \, Z_n^T | \mathcal{O}_n).$$

According to (33) the assertion is obtained by taking again expectations.

In order to simplify our analysis of (34), let the decomposition (32)
be given by (25), i.e.

$$\mathbb{N}^{(2)} := \{\bar{n} + k\cdot(n_S+n_D)+\ell : k\in \mathbb{N}_o, \ \ell \in \{1,\ldots,n_S\}\}$$

$$\mathbb{N}^{(1)} := \mathbb{N}\setminus \mathbb{N}^{(2)} \tag{35}$$

with constant $\bar{n}$, n_S, $n_D \in \mathbb{N}$. As in section 4 we again assume

$$\mathbb{M}_t := \{\bar{n}+k\cdot(n_S+n_D) + t : k \in \mathbb{N}_o\}$$

to be the set of all t-th stochastic steps and all (t-n_S)-th deterministic steps respectively for $t \in \{1,\ldots,n_S+n_D\}$.

We make use of

__Lemma 5.2.__ For every $y\in \mathbb{R}$, where $|y|<1$, $K\in \mathbb{N}_o$ and $t \in \{1,\ldots,n_S+n_D\}$, we obtain

a) $G_1 := \lim\limits_{\mathbb{M}_t \ni N\to\infty} \sum\limits_{M\in \mathbb{N}_{K,N}^{(1)}} y^{N-M} =$

$$= y^t \cdot \frac{1-y^{n_D}}{(1-y)(1-y^{n_S+n_D})} + \frac{1-y^{\max\{0,t-n_S\}}}{1-y}$$

b) $G_2 := \lim\limits_{\mathbb{M}_t \ni N\to\infty} \sum\limits_{M\in \mathbb{N}_{K,N}^{(2)}} y^{N-M} = \dfrac{1}{1-y} - G_1$

__Proof.__ Since $\lim\limits_{N\to\infty} y^{N-M} = 0$ for every $M\in \mathbb{N}$, we may assume $K\in \mathbb{N}^{(1)}$ and $K+1\in \mathbb{N}^{(2)}$. If $K \le N \in \mathbb{M}_t$ and $M \in \mathbb{N}_{K,M}^{(1)}$, then

$$N = K + n(n_S+n_D) + t$$

$$M = K + m(n_S+n_D) + j, \text{ where}$$

$n \in \mathbb{N}_o$, $m\in\{0,\ldots,n\}$, $j\in\{n_S+1,\ldots,n_S+n_D\}$ and $j \le t$ for $m = n$.

Hence,

$$\sum\limits_{M\in \mathbb{N}_{K,N}^{(1)}} y^{N-M} = y^t \cdot \sum\limits_{m=0}^{n-1} (y^{n_S+n_D})^{n-m} \sum\limits_{i=1}^{n_D} y^{-(n_S+i)} +$$

$$+ \sum\limits_{i=1}^{\max\{0,t-n_S\}} y^{t-(n_S+i)} =$$

$$= y^t \sum\limits_{r=0}^{n-1} (y^{n_S+n_D})^r \cdot \sum\limits_{s=0}^{n_D-1} y^s + \sum\limits_{s=0}^{\max\{0,t-n_S\}-1} y^s =$$

$$= y^t \cdot \frac{1-(y^{n_S+n_D})^n}{1-y^{n_S+n_D}} \cdot \frac{1-y^{n_D}}{1-y} + \frac{1-y^{\max\{0,t-n_S\}}}{1-y}.$$

The limiting process $n \to \infty$ yields assertion a). The formula given in b) results from the following relation:

$$G_1 + G_2 = \lim_{\mathbb{M}_t \ni N \to \infty} \sum_{M=K}^{N} y^{N-M} = \frac{1}{1-y} \;.$$

Lemma 5.3. Let $U^{(0)}, C^{(1)}, C^{(2)}$ be symmetric matrices. For $n \in \mathbb{N}_o$ we define

$$U^{(n+1)} := (I-rH)U^{(n)} \cdot (I-rH) + \begin{cases} C^{(1)}, & n \in \mathbb{N}^{(1)} \\ C^{(2)}, & n \in \mathbb{N}^{(2)} \end{cases} \;.$$

Thus, for every $t \in (1, \ldots, n_S + n_D)$ there exists the limit

$$U_t := \lim_{\mathbb{M}_t \ni n \to \infty} U^{(n+1)}$$

which is calculated using the following formula:

$$U_t = P \; \hat{U}_t \; P^T \quad \text{with}$$

$$\hat{U}_t := \left(\frac{1}{1-P_{ij}} \{ \hat{c}_{ij}^{(2)} + [\hat{c}_{ij}^{(1)} - \hat{c}_{ij}^{(2)}] \cdot \right.$$

$$\left. \cdot [P_{ij}^t \cdot \frac{1 - P_{ij}^{n_D}}{1 - P_{ij}^{n_S+n_D}} + 1 - P_{ij}^{\max(0, t-n_S)}] \} \right)_{1 \le i,j \le \nu} ,$$

$$\hat{C}^{(k)} = (\hat{c}_{ij}^{(k)})_{ij} := P^T \cdot C^{(k)} \cdot P \qquad \text{for } k=1,2, \text{ and}$$

$$P_{ij} := (1-rh_i)(1-rh_j) \qquad \text{for } 1 \le i,j \le \nu.$$

Furthermore we have

$$||U_t|| \le \frac{\sqrt{\nu}}{2\alpha r - \beta^2 r^2} (||C^{(1)}|| + ||C^{(2)}||).$$

Proof. By complete induction it is shown that for every n

$$U^{(n+1)} = B^{n+1} \cdot U^{(o)} \cdot B^{n+1} + \sum_{k=1}^{2} \sum_{m \in \mathbb{N}_{o,n}^{(k)}} B^{n-m} \cdot C^{(k)} \cdot B^{n-m},$$

where $B := I-rH$. Following transformation: $\hat{U}^{(n)} := P^T U^{(n)} P$, $\hat{B} := P^T B P$, we obtain for the component (i,j) of the matrices:

$$\hat{u}_{ij}^{(n+1)} = P_{ij}^{n+1} \hat{u}_{ij}^{(o)} + \sum_{k=1}^{2} \hat{c}_{ij}^{(k)} \sum_{m \in \mathbb{N}_{o,n}^{(k)}} P_{ij}^{n-m} \;.$$

Since $|p_{ij}| < 1$, Lemma 5.2 yields

$$\lim_{\mathbb{M}_t \ni n \to \infty} \hat{u}_{ij}^{(n+1)} - [\hat{c}_{ij}^{(1)} - \hat{c}_{ij}^{(2)}] \cdot [p_{ij}^t \cdot \frac{1-p_{ij}^{n_D}}{(1-p_{ij})(1-p_{ij}^{n_S+n_D})} +$$

$$+ \frac{1-p_{ij}^{\max(0,t-n_S)}}{1-p_{ij}}] + \frac{\hat{c}_{ij}^{(2)}}{1-p_{ij}} =: \hat{u}_{ij}.$$

Furthermore, we have

$$|\hat{u}_{ij}| \le \sum_{k=1}^{2} |\hat{c}_{ij}^{(k)}| \lim_{\mathbb{M}_t \ni n \to \infty} \sum_{m \in \mathbb{N}_{o,n}^{(k)}} |p_{ij}|^{n-m} \le$$

$$\le \max \{|\hat{c}_{ij}^{(1)}|, |\hat{c}_{ij}^{(2)}|\} \cdot \frac{1}{1-|p_{ij}|}.$$

Since $|p_{ij}| \le 1-2\alpha r+\beta^2 \cdot r^2$, we obtain

$$||U_t|| - ||(\hat{u}_{ij})|| \le \frac{\sqrt{\nu}}{2\alpha r-\beta^2 r^2} (||\hat{c}^{(1)}|| + ||\hat{c}^{(2)}||)$$

$$- \frac{\sqrt{\nu}}{2\alpha r-\beta^2 r^2} (||c^{(1)}|| + ||c^{(2)}||).$$

On the basis of this preliminary work we are now in a position to prove the following limit theorem.

Theorem 5.2. For every given $\epsilon > 0$ there is an index n_o such that

$$E \frac{1}{r} \Delta_{n+1} \Delta_{n+1}^T \le V_t + \epsilon \cdot I.$$

for every $t \in \{1,\ldots,n_S+n_D\}$ and $n \ge n_o, n \in \mathbb{M}_t$.

Matrix V_t is - independently of ϵ - given by

$$V_t - P \cdot \hat{V}_t \cdot P^T \text{ with} \tag{36.1}$$

$$\hat{V}_t := \left(\frac{r}{1-p_{ij}} \cdot (w_{ij}^{(2)} + \delta_{ij} \cdot K^{(2)} + [\hat{w}_{ij}^{(1)} - \hat{w}_{ij}^{(2)} + \delta_{ij} \cdot (K^{(1)} - K^{(2)})] \cdot \right.$$

$$\left. \cdot [p_{ij}^t \cdot \frac{1-p_{ij}^{n_D}}{1-p_{ij}^{n_S+n_D}} + 1 - p_{ij}^{\max(0,t-n_S)}]) \right)_{1 \le i,j \le \nu}, \tag{36.2}$$

$$\hat{W}^{(k)} - (\hat{w}_{ij}^{(k)})_{ij} := - P^T \cdot W^{(k)} \cdot P, \tag{36.3}$$

$$K^{(k)} := u_k \sqrt{L} + v_k L \quad \text{for } k=1,2, \text{ where } L := r \frac{\sigma^2}{2\alpha - (\beta^2 + \gamma^2)r} , \quad (36.4)$$

$$P_{ij} := (1 - rh_i) \cdot (1 - rh_j) \quad (34.5)$$

$$\text{and } \delta_{ij} := \begin{cases} 1, & i=j \\ 0, & \text{otherwise} \end{cases} \quad \text{for } 1 \le i, j \le \nu .$$

<u>Proof.</u> Let $\epsilon > 0$ and $t \in \{1, \ldots, n_S + n_D\}$.

According to Theorem 5.1 there is an index n_1 with

$$u_k E ||\Delta_n|| + v_k E ||\Delta_n||^2 \le K^{(k)} + \frac{\epsilon}{4\sqrt{\nu}} (2\alpha - r\beta^2) \quad (i)$$

for every $k \in \{1,2\}$ and $n \ge n_1$. For $e \in \mathbb{R}$ and $k \in \{1,2\}$ and $n \ge n_1$ we now take

$$C^{(k)}(e) := r \cdot (W^{(k)} + (K^{(k)} + e) \cdot I)$$

$$U^{(n_1)}(e) := E \frac{1}{r} \Delta_{n_1} \Delta_{n_1}^T$$

$$U^{(n+1)}(e) := B \, U^{(n)}(e) \, B + \begin{cases} C^{(1)}(e), & n \in \mathbb{N}^{(1)} \\ C^{(2)}(e), & n \in \mathbb{N}^{(2)} \end{cases} ,$$

where $B := I - rH$. According to Lemma 5.1 and (i) and by complete induction on n we obtain

$$E \frac{1}{r} \Delta_n \Delta_n^T \le U^{(n)}(\epsilon_1) \quad \text{for all } n \ge n_1, \quad (ii)$$

where $\epsilon_1 := \frac{\epsilon}{4\sqrt{\nu}} (2\alpha - r\beta^2)$.

Lemma 5.3 states that for every $e \in \mathbb{R}$ there is the limit

$$U_t(e) := \lim_{\mathbb{M}_t \ni n \to \infty} U^{(n+1)}(e) .$$

Thus, there is an index $n_0 \ge n_1$ with

$$U^{(n+1)}(\epsilon_1) \le U_t(\epsilon_1) + \frac{\epsilon}{2} \cdot I \quad (iii)$$

for all n where $n_0 \le n \in \mathbb{M}_t$.

For every $n \ge n_1$ we have

$$U^{(n+1)}(\epsilon_1) - U^{(n+1)}(0) = B \cdot (U^{(n)}(\epsilon_1) - U^{(n)}(0)) \cdot B + r \cdot \epsilon_1 \cdot I,$$

from which - according to Lemma 5.3 - we obtain

$$||U_t(\epsilon_1) - U_t(0)|| \le \frac{\sqrt{\nu}}{2\alpha r - \beta^2 r^2} 2r\epsilon_1 = \frac{\epsilon}{2} ,$$

hence

$$U_t(\epsilon_1) \le U_t(0) + \frac{\epsilon}{2} \cdot I. \quad (iv)$$

Because of (ii), (iii) and (iv) we now have

$$E \frac{1}{r} \Delta_{n+1} \Delta_{n+1}^T \leq U_t(0) + \epsilon \cdot I$$

for every $n \geq n_o$ with $n \in \mathbb{M}_t$.

According to Lemma 5.3 $U_t(0)$ exactly coincides with matrix V_t as defined in Theorem 5.2.

Theorem 5.2 provides a proposition on the limiting behavior of $(\frac{1}{\sqrt{r}} \cdot \Delta_n)_n$ for arbitrary step size values r satisfying (30.2). For very small values of r we obtain

Theorem 5.3. For every $\epsilon > 0$ there are (small) step size values of $r > 0$, and there is an index n_o such that

$$E \frac{1}{r} \Delta_n \Delta_n^T \leq V + \epsilon \cdot I.$$

for all $n \geq n_o$. Independently of ϵ and r matrix V is given by the matrix equation

$$H \cdot V + V \cdot H = \frac{n_D}{n_S + n_D} \cdot W^{(1)} + \frac{n_S}{n_S + n_D} \cdot W^{(2)}. \tag{37}$$

Proof. Let V_t und $\hat{V}_t$ be the matrices according to (36) for $t \in \{1, \ldots, n_S + n_D\}$.

Clearly we obtain

$$\lim_{r \to o} \hat{V}_t = (\frac{1}{h_i + h_j} \cdot (\hat{w}_{ij}^{(2)} + [\hat{w}_{ij}^{(1)} - \hat{w}_{ij}^{(2)}] \cdot \frac{n_D}{n_S + n_D}))_{i,j} =: \hat{V},$$

hence

$$\lim_{r \to o} V_t = P \hat{V} P^T = V, \tag{i}$$

with V as defined by (37).

Assume arbitrary values of $\epsilon > 0$. Due to (i) there is a step size $r > 0$ with

$$||V_t - V|| < \frac{\epsilon}{2} \text{ for every } t \in \{1, \ldots, n_S + n_D\}. \tag{ii}$$

According to Theorem 5.2 there is an index n_o for these values ϵ and r such that

$$E \frac{1}{r} \Delta_{n+1} \Delta_{n+1}^{T} \leq V_t + \frac{\epsilon}{2} \cdot I \qquad \text{(iii)}$$

for all $t \in \{1, \ldots, n_S + n_D\}$, $n_0 \leq n \in M_t$.

(ii) und (iii), however, yield

$$E \frac{1}{r} \Delta_{n+1} \Delta_{n+1}^{T} \leq V + \epsilon \cdot I$$

for all $n \geq n_0$.

Matrix $\tilde{W} := \frac{n_D}{n_S + n_D} W^{(1)} + \frac{n_S}{n_S + n_D} W^{(2)}$ as defined on the right side of

equation (37) may be interpreted as "mean covariance matrix" of the

estimation error $Z(\omega)$ at x^*.

6. A special representation of the sequence of estimation errors $(Z_n)_n$

Many applications (see, for example, section 6.1) use the following

representation of the sequence of estimation errors $(Z_n)_n$ in (2):

Let us again start from a given disjoint decomposition

$$\mathbb{N} = \mathbb{N}^{(1)} \cup \ldots \cup \mathbb{N}^{(\kappa)} \qquad (38)$$

of $\mathbb{N}$, where $|\mathbb{N}^{(k)}| = \infty$ for $k = 1, \ldots, \kappa$.

Further, we assume for every $k \in \{1, \ldots, \kappa\}$ natural numbers ν_k, N_k,

constants $s_k, t_k \geq 0$, a continuous function $\psi^{(k)}: \mathbb{R}^\nu \times \mathbb{R}^{\nu_k} \longrightarrow \mathbb{R}^\nu$ and

a random vector $\Pi^{(k)}: \Omega \longrightarrow \mathbb{R}^{\nu_k}$ with the following properties:

$$E \psi^{(k)}(x, \Pi^{(k)}) = 0 \qquad \text{for all } x \in \mathbb{R}^\nu, \qquad (39.1)$$

$$E ||\psi^{(k)}(x, \Pi^{(k)})||^2 \leq s_k^2 + t_k^2 \cdot ||x - x^*||^2 \qquad \text{for all } x \in \mathbb{R}^\nu, \qquad (39.2)$$

$$\phi^{(k)}(x) := E \psi^{(k)}(x, \Pi^{(k)}) \cdot \psi^{(k)}(x, \Pi^{(k)})^T \text{ continuous in } x^*, \qquad (39.3)$$

For every $n \in \mathbb{N}^{(k)}$ the error of estimation is given by

$$Z_n = \frac{1}{N_k} \sum_{i=1}^{N_k} \psi^{(k)}(X_n, \Pi_{n,i}^{(k)}), \qquad (39.4)$$

where $\Pi_{n,1}^{(k)}, \ldots, \Pi_{n,N_k}^{(k)}$ are independent random vectors distributed like

$\Pi^{(k)}$ which are furthermore supposed to be independent of σ-algebra

$\mathcal{O}_n$ (especially of X_n).

Because of (39.3) the function

$$\varphi^{(k)}(x) \colon \; - \; E||\psi^{(k)}(x,\Pi^{(k)})||^2 \; - \; trace(\phi^{(k)}(x)) \qquad\qquad (39.5)$$

is continuous in x^*.

According to (39.4), Z_n is the arithmetic mean of N_k independent

realizations of the random vector $\psi^{(k)}(X_n,\Pi^{(k)})$ for a given X_n and

$n \in \mathbb{N}^{(k)}$. Additional properties of $(Z_n)_n$ are given in

__Theorem 6.1.__ For every $k \in \{1,\ldots,\kappa\}$ and $n \in \mathbb{N}^{(k)}$ the following holds

true:

a) $E(Z_n|\mathcal{O}\!l_n) \equiv 0$ a.s.

b) $E(||Z_n||^2) \leq \frac{1}{N_k} \cdot (s_k^2 + t_k^2 \; E||\Delta_n||^2)$

c) $E(||Z_n||^2|\mathcal{O}\!l_n) \; - \; \frac{1}{N_k} \; \varphi^{(k)}(X_n)$ a.s.

d) $E(Z_n \; Z_n{}^T|\mathcal{O}\!l_n) \; - \; \frac{1}{N_k} \; \phi^{(k)}(X_n)$ a.s.

If

$$\lim_{n \to \infty} ||\Delta_n|| - 0 \quad \text{a.s.},$$

then

e) $\lim\limits_{\mathbb{N}^{(k)} \ni n \to \infty} \quad E(||Z_n||^2|\mathcal{O}\!l_n) \; - \; \frac{1}{N_k} \; \varphi^{(k)}(x^*)$ a.s.

f) $\lim\limits_{\mathbb{N}^{(k)} \ni n \to \infty} \quad E(Z_n \; Z_n{}^T|\mathcal{O}\!l_n) \; - \; \frac{1}{N_k} \; \phi^{(k)}(x^*)$ a.s.

for every $k \in \{1,\ldots,\kappa\}$.

__Proof.__ Assertion a) follows directly from (39.1) and (39.4). Due to

(39.4) we have for $k \in \{1,\ldots,\kappa\}$ and $n \in \mathbb{N}^{(k)}$

$$Z_n \; Z_n{}^T \; - \; \frac{1}{N_k^2} \; \sum_{i,j-1}^{N_k} \; \psi^{(k)}(X_n,\Pi_{n,i}^{(k)}) \cdot \psi^{(k)}(X_n,\Pi_{n,j}^{(k)})^T,$$

from which assertion d) follows according to (39).

Via a trace function equation c) follows from d), and inequality b)

results from c) and (39.2).

If $\lim\limits_{n \to \infty} ||\Delta_n|| - 0$ a.s., then (39.3) and (39.5) yield for every

$k \in \{1,\ldots,\kappa\}$

$$\lim_{n \to \infty} \varphi^{(k)}(X_n) = \varphi^{(k)}(x^*) \quad \text{a.s.}$$

$$\lim_{n \to \infty} \phi^{(k)}(X_n) = \phi^{(k)}(x^*) \quad \text{a.s.}$$

from which proposition e) and f) follows according to c) and d).

According to Theorem 6.1 we may state that

1. Conditions (4) are satisfied by

$$\sigma^2: = \max_{1 \le k \le \kappa} \frac{s_k^2}{N_k} \ , \ \gamma^2: = \max_{1 \le k \le \kappa} \frac{t_k^2}{N_k} \ .$$

2. Condition (10) of Theorem 3.2 is satisfied by

$$\sigma_k^2: = \frac{s_k^2}{N_k} \ , \ k=1,\ldots,\kappa.$$

3. Premise (14.1) of Theorem 3.3 is satisfied by

$$W^{(k)}: = \frac{1}{N_k} \phi^{(k)}(x^*), \ k=1,\ldots,\kappa.$$

Theorem 4.5 requires the existence of a limit of unconditional expectations values of $Z_n Z_n^T$. A pertinent proposition is provided by

<u>Theorem 6.2.</u> If $\sum\limits_{n \in \mathbb{N}} E||\Delta_n||^2 < \infty$, then

e') $\lim\limits_{\mathbb{N}^{(k)} \ni n \to \infty} E||Z_n||^2 = \frac{1}{N_k} \varphi^{(k)}(x^*)$

f') $\lim\limits_{\mathbb{N}^{(k)} \ni n \to \infty} E Z_n Z_n^T = \frac{1}{N_k} \phi^{(k)}(x^*).$

for every $k \in (1,\ldots,\kappa)$:

<u>Proof.</u> Choose arbitrary values of $k \in (1,\ldots,\kappa)$. Theorem 6.1c) and (39.2) yield for every $n \in \mathbb{N}^{(k)}$

$$||E(Z_n Z_n^T|\mathcal{O}_n)|| \le E(||Z_n||^2|\mathcal{O}_n) \le \gamma^{(k)}$$

where $\gamma^{(k)}: = \frac{1}{N_k} \cdot (s_k^2 + t_k^2 \cdot \sum\limits_{m \in \mathbb{N}^{(k)}} ||\Delta_m||^2).$

$\underset{n\in\mathbb{N}}{\Sigma}\ E||\Delta_n||^2 < \infty$ yields

$E\ Y^{(k)} < \infty$ and $\lim_{n\to\infty} ||\Delta_n|| = 0$ almost sure.

Hence, according to Lebesgue's theorem and Theorem 6.1 e) and f) we

have

$$\lim_{\mathbb{N}^{(k)}\ni n\to\infty} E||Z_n||^2 = E\lim_{\mathbb{N}^{(k)}\ni n\to\infty} E(||Z_n||^2|\mathcal{O}_n) = \frac{1}{N_k}\varphi^{(k)}(x^*)$$

and

$$\lim_{\mathbb{N}^{(k)}\ni n\to\infty} E\ Z_n\ Z_n^T = E\lim_{\mathbb{N}^{(k)}\ni n\to\infty} E(Z_n\ Z_n^T|\mathcal{O}_n) = \frac{1}{N_k}\phi^{(k)}(x^*).$$

Because of Theorem 4.3, in section 4.2 the assumption $\underset{n\in\mathbb{N}}{\Sigma}\ E||\Delta_n||^2 < +\infty$

is fulfilled. Theorem 6.2 confirms the existence of the matrix

$$W^{(2)}: = \lim_{\mathbb{N}^{(2)}\ni n\to\infty} E\ Z_n\ Z_n^T$$

according to Theorem 4.5, and we have

$$W^{(2)} = \frac{1}{N_2}\phi^{(2)}(x^*)\ .$$

It remains to be clarified under which conditions assumption (14.2)

of Theorem 3.3 is valid. For this purpose we refer to

<u>Lemma 6.1.</u> Fo every $k\in(1,\ldots,\kappa)$, $n\in\mathbb{N}^{(k)}$ and $R>0$ we have

$$E(||Z_n||^2\cdot I_{(||Z_n||^2>R)}|\mathcal{O}_n) \leq \hat{\varphi}^{(k)}(X_n,R) + \frac{N_k-1}{R}\varphi^{(k)}(X_n)^2,$$

defining for $x\in\mathbb{R}^\nu$, $R>0$

$$\hat{\varphi}^{(k)}(x,R): = E(||\psi^{(k)}(x,\Pi^{(k)})||^2\cdot I_{(||\psi^{(k)}(x,\Pi^{(k)})||^2 > R)}).$$

<u>Proof.</u> Choose fixed values of k,n,R as in Lemma 6.1 above.

According to (39.4) we obtain - for $a_i: = \psi^{(k)}(X_n,\Pi_{n,i}^{(k)})$

$$||Z_n||^2 - \frac{1}{N_k^2}\sum_{i,j=1}^{N_k} a_i^T\cdot a_j \leq \frac{1}{N_k^2}\sum_{i,j=1}^{N_k} ||a_i||\cdot||a_j||$$

$$\leq \frac{1}{N_k}\sum_{i=1}^{N_k} ||a_i||^2, \text{ hence}$$

$$||Z_n||^2 \cdot I_{(||Z_n||^2 > R)} \leq \frac{1}{N_k} \sum_{i=1}^{N_k} ||a_i||^2 \cdot I_{(\frac{1}{N_k} \sum_{j=1}^{N_k} ||a_j||^2 > R)}$$

$$\leq \frac{1}{N_k} \sum_{i,j=1}^{N_k} ||a_i||^2 \cdot I_{(||a_j||^2 > R)} .$$

Hence, taking expectations yields

$$E(||Z_n||^2 \cdot I_{(||Z_n||^2 > R)} |\mathcal{O}_n) \leq \frac{1}{N_k} (\sum_{i=1}^{N_k} E(||a_i||^2 \cdot I_{(||a_i||^2 > R)} |\mathcal{O}_n) +$$

$$+ \sum_{i \neq j} E (||a_i||^2 |\mathcal{O}_n) \cdot \mathbb{P}(||a_j||^2 > R |\mathcal{O}_n)).$$

From this and from the relations

$$E(||a_i||^2 |\mathcal{O}_n) - \varphi^{(k)}(X_n),$$

$$\mathbb{P}(||a_j||^2 > R |\mathcal{O}_n) \leq \frac{1}{R} E (||a_j||^2 |\mathcal{O}_n),$$

$$E(||a_i||^2 \cdot I_{(||a_i||^2 > R)} |\mathcal{O}_n) - \hat{\varphi}^{(k)}(X_n, R)$$

we obtain the assertion of this lemma.

Now we are able to answer the above-mentioned question:

<u>Theorem 6.3.</u> Given that $\lim_{n \to \infty} ||\Delta_n|| - 0$ almost sure, and that there

is a number $\bar{R} > 0$ such that for the function $\hat{\varphi}^{(k)}$ given in Lemma 6.1

we have:

$$\hat{\varphi}^{(k)}(\cdot, R) \text{ is continuous in } x^* \text{ for every } 1 \leq k \leq \kappa \text{ and } R \geq \bar{R}. \qquad (39.6)$$

Then, assumption (14.2) is fulfilled, i.e.

$$\lim_{n \to \infty} E(||Z_n||^2 \cdot I_{(||Z_n||^2 > t \cdot n)} |\mathcal{O}_n) - \text{ almost sure for every } t > 0.$$

<u>Proof.</u> Given $t > 0$ and arbitrary values of $k \in \{1, \ldots, \kappa\}$. As $X_n \to x^*$

almost sure, and because of (39.5) we find

$$\lim_{n \to \infty} \frac{N_k - 1}{t \cdot n} \varphi^{(k)}(X_n)^2 - 0 \quad \text{a.s.}$$

Thus, according to Lemma 6.1., we only have to verify the following

relation

$$\lim_{n \to \infty} \hat{\varphi}^{(k)}(X_n, t \cdot n) - 0 \quad \text{almost sure.} \qquad (*)$$

For this purpose we choose $\epsilon > 0$ and $\omega \in \Omega$ with $x^* = \lim_{n \to \infty} X_n(\omega)$. For every $x \in \mathbb{R}^\nu$ $\hat{\varphi}^{(k)}(x, \cdot)$ is monotone decreasing and $\lim_{R \to \infty} \hat{\varphi}^{(k)}(x, R) = 0$. There is, therefore, an index $n_1 \geq \bar{R}/t$ with

$$\hat{\varphi}^{(k)}(x^*, t \cdot n_1) < \frac{\epsilon}{2} .\tag{i}$$

Since $t \cdot n_1 \geq \bar{R}$ and because of (39.6) we now obtain an index $n_0 \geq n_1$ with

$$\hat{\varphi}^{(k)}(X_n(\omega), t \cdot n_1) < \hat{\varphi}^{(k)}(x^*, t \cdot n_1) + \frac{\epsilon}{2} \tag{ii}$$

for all $n \geq n_0$. (i) and (ii) yield for all $n \geq n_0$

$$\hat{\varphi}^{(k)}(X_n(\omega), t \cdot n) \leq \hat{\varphi}^{(k)}(X_n(\omega), t \cdot n_1) \leq \epsilon,$$

hence, (*) is fulfilled.

6.1. Example

A large group of stochastic optimization problems involves the following problem (cf. [7]):

The quantity to be found is the minimum point $x^* \in \mathbb{R}^\nu$ of a function $F: \mathbb{R}^\nu \to \mathbb{R}$, given by

$$F(x) = c^T \cdot x + Eu(A(\omega) \cdot x - b(\omega)),\tag{40}$$

where $c \in \mathbb{R}^\nu$, $(A,b): \Omega \to \mathbb{R}^{m \times (\nu+1)}$ is a random matrix and u: $\mathbb{R}^m \to \mathbb{R}$ is a convex and twice continuously differentiable function. Under weak conditions we obtain

$$G(x): = c + EA(\omega)^T \cdot \nabla u(A(\omega) \cdot x - b(\omega)),\tag{41}$$

where $G(x)$ is the gradient of $F(x)$, and x^* fulfils the necessary, and here also sufficient, condition (1), i.e. in this case

$$c + EA^T \nabla u(Ax^* - b) = 0.\tag{42}$$

In (39) we choose therefore for $k = 1, \ldots, \kappa$

$$\Pi^{(k)} := \Pi := (A,b): \Omega \to \mathbb{R}^{m \times (\nu+1)}\tag{43.1}$$

independently of k. In equation (2.2) we use the following stochastic gradient in step $n \in \mathbb{N}^{(k)}$, $k \in \{1, \ldots, \kappa\}$

$$\hat{G}_n(x) =: c + \frac{1}{N_k} \sum_{i=1}^{N_k} U^{(k)}(x, \Pi_{n,i}),\tag{43.2}$$

where $\Pi_{n,1}, \ldots, \Pi_{n,N_k}$ denote random matrices which are independent of each other as well as of $\mathcal{O}\!\ell_n$ and which have a distribution being equal to $\Pi=(A,b)$. Moreover to the function $U^{(k)} : \mathbb{R}^\nu \times \mathbb{R}^{m \times (\nu+1)} \longrightarrow \mathbb{R}^\nu$ we assume

$$E \; U^{(k)}(x,\Pi) = E \; A^T \nabla u(Ax-b), \quad x \in \mathbb{R}^\nu. \tag{43.3}$$

Assumption (43.3) is, for example, satisfied by the functions

$$U^{(k)}(x,\Pi) := U_o(x,\Pi) := A^T \nabla u(Ax-b) \tag{44.1}$$

or

$$U^{(k)}(x,\Pi) := E \; A^T \nabla u(Ax-b) \tag{44.2}$$

Since $Z_n = \hat{G}_n(X_n) - G(X_n)$, and because of (41) and (43.2), function $\psi^{(k)}$ from (39) is given by

$$\psi^{(k)}(x,\Pi) := U^{(k)}(x,\Pi) - E \; A^T \nabla u(Ax-b). \tag{45}$$

(43.3) confirms (39.1), and conditions (39.2) and (39.3) mean that for all $k \in \{1, \ldots, \kappa\}$

$$E||U^{(k)}(x,\Pi) - E \; A^T \nabla u(Ax-b)||^2 \leq s_k^2 + t_k^2 ||x-x^*||^2 \tag{43.4}$$

for all $x \in \mathbb{R}^\nu$ and

$$\phi^{(k)}(x) := E(U^{(k)}(x,\Pi) - E \; A^T \nabla u(Ax-b)) \cdot$$
$$\cdot (U^{(k)}(x,\Pi) - E \; A^T \nabla u(Ax-b))^T \tag{43.5}$$

is continuous in x^*.

According to (43.3) we further have

$$\phi^{(k)}(x) = E \; U^{(k)}(x,\Pi) \cdot U^{(k)}(x,\Pi)^T$$
$$- (E \; A^T \nabla u(Ax-b)) \cdot (E \; A^T \nabla u(Ax-b))^T.$$

Thus, on the basis of (42) the covariance matrices of Theorems 6.1 and 6.2 are given by

$$\frac{1}{N_k} \phi^{(k)}(x^*) = \frac{1}{N_k} (E \; U^{(k)}(x^*,\Pi) \cdot U^{(k)}(x^*,\Pi)^T - c \; c^T). \tag{46}$$

for $k=1, \ldots, \kappa$.

For x^* the Jacobian matrix H of G is obviously defined by

$$H = E \; A^T(\omega) \cdot \nabla^2 u(A(\omega)x^*-b(\omega)) \cdot A(\omega) \tag{47}$$

The covariance matrix V or V_t given in Theorems 3.3, 4.5, 5.2 and 5.3 can be calculated - at least theoretically - by using (46) and (47).

Annex A. On special sequences, series, and products.

This annex summarizes several equations and inequations on sequences and series which are needed in the previous sections. Some of these propositions are to be found in the pertinent literature (cf.[6],[13],[14]). For notational simplification we take:

$$\mathcal{P}: - ((a_n)_{n \in \mathbb{N}} \qquad : a_n \geq 0 \text{ for all } n \in \mathbb{N}),$$

$$\mathcal{Q}: - ((a_{m,n})_{\substack{n \in \mathbb{N} \\ m \in \{1,\ldots,n\}}} : \begin{array}{l} \text{i) there is an } n_o: a_{m,n} > 0 \text{ for all } n_o \leq m \leq n \\ \text{ii) } \lim_{n \to \infty} a_{m,n} = 0 \text{ for all } m \in \mathbb{N} \end{array}).$$

Lemma A.1. (O. Toeplitz, cf. [6] Paragraph 8, Theorem 5).

Given $n_o \in \mathbb{N}$, non-negative numbers K_1, K_2, L_1, L_2, a, b and $(b_m)_m, (\beta_m)_m \in \mathcal{P}$ $(a_{m,n})_{m \leq n}, (\alpha_{m,n})_{m \leq n} \in \mathcal{Q}$, where

i) $\quad K_1 a_{m,n} \leq \alpha_{m,n} \leq K_2 a_{m,n}$

$\quad L_1 b_m \leq \beta_m \leq L_2 b_m$

$\quad$ for all $m, n \in \mathbb{N}$ with $n_o \leq m \leq n$,

ii) $\lim_{n \to \infty} \sum_{m=1}^{n} a_{m,n} = a,$

iii) $\lim_{m \to \infty} b_m = b.$

Then we have

a) $\operatorname{limextr}_{n \to \infty} \sum_{m=k}^{n} \alpha_{m,n} \beta_m \in [K_1 L_1 ab, K_2 L_2 ab]$

and specially

b) $\lim_{n \to \infty} \sum_{m=k}^{n} a_{m,n} \cdot b_m = ab.$

for all $k \in \mathbb{N}$, where limextr denotes either limsup or liminf.

In this context we are interested in the so-called equivalent double sequences.

Thus, we establish

Definition A.1. Let $(a_{m,n})_{m \leq n}$ and $(\alpha_{m,n})_{m \leq n}$ be arbitrarily chosen real double sequences.

a) $(a_{m,n})_{m\le n}$ is _not greater_ than $(\alpha_{m,n})_{m\le n}$ - written as $a_{m,n} \ll \alpha_{m,n}$ -
   if for every $\epsilon > 0$ there is an $n_0 \in \mathbb{N}$ such that $0 < a_{m,n} \le (1+\epsilon)\alpha_{m,n}$
   for all $m,n \in \mathbb{N}$ with $n_0 \le m \le n$.

b) $(a_{m,n})_{m\le n}$ is _equivalent_ to $(\alpha_{m,n})_{m\le n}$ - written as $a_{m,n} \sim \alpha_{m,n}$, if
   $a_{m,n} \ll \alpha_{m,n}$ and $\alpha_{m,n} \ll a_{m,n}$.

The relation "$\ll$" is an order and "$\sim$" is an equivalence relation on $\mathfrak{D}$.

Properties of equivalent sequences are given in the following.

Lemma A.2. Let $(a_{m,n})_{m\le n}$, $(b_{m,n})_{m\le n}$, $(\alpha_{m,n})_{m\le n}$ and $(\beta_{m,n})_{m\le n}$ be double
sequences of $\mathfrak{D}$, and $(b_m)_m$ a sequence in $\mathcal{P}$.

a) If $a_{m,n} \ll \alpha_{m,n}$ and $b_{m,n} \ll \beta_{m,n}$, then it follows that $a_{m,n} + b_{m,n} \ll$
   $\alpha_{m,n} + \beta_{m,n}$, $a_{m,n} \cdot b_{m,n} \ll \alpha_{m,n} \cdot \beta_{m,n}$ and $a_{m,n}{}^s \ll \alpha_{m,n}{}^s$ for all $s>1$.

b) If $a_{m,n} \sim \alpha_{m,n}$ and $b_{m,n} \sim \beta_{m,n}$, then it follows that $a_{m,n} + b_{m,n} \sim$
   $\alpha_{m,n} + \beta_{m,n}$, $a_{m,n} \cdot b_{m,n} \sim \alpha_{m,n} \cdot \beta_{m,n}$ and $a_{m,n}{}^s \sim \alpha_{m,n}{}^s$ for all $s>0$.

c) Let a,b be non-negative numbers with

   i) $\lim\limits_{n \to \infty} \sum\limits_{m=1}^{n} a_{m,n} = a$

   ii) $\lim\limits_{m \to \infty} b_m = b$.

Then, we find

   c1) If $a_{m,n} \ll \alpha_{m,n}$, then it follows that
       $$a \cdot b \le \liminf_{n \to \infty} \sum_{m=k}^{n} \alpha_{m,n} \cdot b_m,$$

   c2) If $\alpha_{m,m} \ll a_{m,n}$, then it follows that
       $$\limsup_{n \to \infty} \sum_{m=k}^{n} \alpha_{m,n} \cdot b_m \le ab,$$

   c3) If $\alpha_{m,n} \sim a_{m,n}$, then it follows that
       $$\lim_{n \to \infty} \sum_{m=k}^{n} \alpha_{m,n} \cdot b_m = ab.$$

   for all $k \in \mathbb{N}$.

Proof. a) and b) directly result from the definition of the relations "≪" and "~". The propositions in c) follow from Lemma A.1.

Propositions on a special sequence of $\mathcal{D}$ are provided by

Lemma A.3. Assume $(a_n)_n \in \mathcal{P}$ with $a_n < 1$ for all $n \in \mathbb{N}$ and $\sum_{n \in \mathbb{N}} a_n = \infty$. Take

$$b_{m,n} := \begin{cases} \prod_{j=m+1}^{n} (1-a_j), & \text{for } m < n \\ 1, & \text{for } m = n \end{cases}$$

for $m+1, n \in \mathbb{N}$ with $m \le n$. Hence,

a) $\lim_{n \to \infty} b_{m,n} = 0$ for all $m \in \mathbb{N}_0$, i.e. $(b_{m,n})_{m \le n} \in \mathcal{D}$.

b) For any convergent sequence $(b_m)_m \in \mathcal{P}$ and $k \in \mathbb{N}$ we find

$$\lim_{n \to \infty} \sum_{m=k}^{n} b_{m,n} \cdot a_m \cdot b_m = \lim_{m \to \infty} b_m.$$

c) For $\sum_{n \in \mathbb{N}} a_n^2 < \infty$ we even obtain $b_{m,n} \sim e^{-(a_{m+1}+ \ldots +a_n)}$.

Propositions a) and b) are taken from [13], and c) is a conclusion from Theorem 10, Paragraph 29 in [6]. For the sake of completeness the deduction of these propositions will be repeated in the following:

Proof. Since $\sum_{n \ge 1} a_n = \infty$ and $b_{m,n} \le e^{-(a_{m+1}+ \ldots + a_m)}$ for $m \le n$, (*)

we obtain $\lim_{n \to \infty} b_{m,n} = 0$ for all $m \in \mathbb{N}_0$.

From this and from the relation

$$\sum_{m=k}^{n} b_{m,n} a_m = \sum_{m=k}^{n} b_{m,n}(1-(1-a_m)) = 1-b_{k-1,n} \quad \text{for all } k \le n,$$

follows assertion b) according to Lemma A.1b). Because of (*), in order to verify c) we only have to prove that for every $\epsilon \in (0,1)$ there is an $n_0 \in \mathbb{N}$ such that

$$e^{-(a_{m+1}+ \ldots + a_n)} \leq (1+\epsilon)b_{m,n}$$

for all $m,n \in \mathbb{N}$ with $n_o \leq m \leq n$.

Let $\epsilon \in (0,1)$. Since $\sum_{n\geq 1} a_n^2 < \infty$, there is an index $n_o \in \mathbb{N}$ with the following properties:

$$-2\,a_n^2 \leq \ln(1-a_n^2) \text{ for all } n \geq n_o,$$

$$2 \sum_{n \geq n_o} a_n^2 \leq \ln(1+\epsilon).$$

Hence, we find

$$b_{m,n} \cdot e^{a_{m+1}+ \ldots + a_n} \geq b_{m,n} \cdot (1+a_{m+1}) \cdot \ldots \cdot (1+a_n)$$

$$= (1-a_{m+1}^2) \cdot \ldots \cdot (1-a_n^2) \geq e^{-2(a_{m+1}^2+ \ldots + a_n^2)} \geq \frac{1}{1+\epsilon}$$

for all $n,m \in \mathbb{N}$ with $n_o \leq m \leq n$.

A simple application of Lemma A.3 is given in

<u>Lemma A.4.</u> Let sequences $(a_n)_n$ and $(b_m)_m$ be defined as in Lemma A.3. We further assume $(c_n)_n \in \mathscr{P}$ and $n_o \in \mathbb{N}$, such that

$$c_{n+1} \leq (1-a_n) \cdot c_n + a_n \cdot b_n$$

for all $n \geq n_o$. Hence,

$$\limsup_{n \to \infty} c_n \leq \lim_{m \to \infty} b_m.$$

<u>Proof.</u> By complete induction on n we obtain

$$c_{n+1} \leq b_{n_o-1,n} \cdot c_{n_o} + \sum_{m=n_o}^{n} b_{m,n} \cdot a_m \cdot b_m,$$

for all $n \geq n_o$ with $(b_{m,n})_{m \leq n}$ as defined in Lemma A.3.

Thus, the assertion follows now from Lemma A.3 a), b).

According to Lemma A.3 - for a positive number a - the double sequence

$$\phi_{m,n}(a) := \begin{cases} \prod_{j=m+1}^{n} (1 - \frac{a}{j}), & \text{for } m \leq n \\ 1, & \text{for } m=n \end{cases}$$

is an element of $\mathscr{D}$. Some properties of this sequence are given in the following.

<u>Lemma A.5.</u> Let $a, s > 0$ and $t \in \mathbb{R}$ such that $A := s \cdot a - t > 0$. Hence,

a) $\phi_{m,n}(a) \sim (\frac{m}{n})^a$,

b) $\phi_{m,n}(a)^s \cdot (\frac{n}{m})^t \sim \phi_{m,n}(A)$,

c) $\lim\limits_{n \to \infty} \phi_{m,n}(a)^s \cdot n^t = 0$ for all $m \in \mathbb{N}_o$,

d) $\lim\limits_{n \to \infty} \sum\limits_{m=k}^{n} \phi_{m,n}(a)^s (\frac{n}{m})^t \cdot \frac{1}{m} = \frac{1}{A}$ for all $k \in \mathbb{N}$.

Note that propositions a) and d) (for the case $s=2, t=1$) can be found also in [13].

<u>Proof.</u> Because of

$$e^{-a \cdot (\frac{1}{m+1} + \ldots + \frac{1}{n})} = e^{a(b_m - b_n)} \cdot (\frac{m}{n})^a$$

with the convergent sequence

$$b_k := \frac{1}{1} + \ldots + \frac{1}{k} - \ln(k), \quad k=1, 2, \ldots$$

we obtain assertion a) according to Lemma A.3c). b) follows from a), and b) yields assertion c) according to Lemma A.3a).

The relation given in b), Lemma A.2 c3), and Lemma A.3 b) confirm the validity of the formula given in d).

At this point we have to strengthen the propositions given in Lemma A.5; this requirement is met by the following two lemmas. We start from a given subset $\mathbb{M}$ of $\mathbb{N}$ and a disjoint decomposition $\mathbb{N} = \mathbb{N}^{(1)} \cup \ldots \cup \mathbb{N}^{(\kappa)}$ of $\mathbb{N}$. As in section 1.1 let

$$\mathbb{M}_{m,n} := \{m+1, \ldots, n\} \cap \mathbb{M}$$

be the <u>(m.n)-segment</u> of $\mathbb{M}$ for $m \leq n$.

We assume existing positive limits

$$q := \lim\limits_{n \to \infty} \frac{|\mathbb{M}_{o,n}|}{n},$$

$$q_k := \lim\limits_{n \to \infty} \frac{|\mathbb{N}_{o,n}^{(k)}|}{n} \quad \text{for } k=1, \ldots, \kappa.$$

<u>Lemma A.6.</u>

a) $\frac{m}{n} \sim \frac{|\mathbb{M}_{o,m}|}{|\mathbb{M}_{o,n}|}$.

b) Let $(a_{m,n})_{m\leq n} \in \mathfrak{A}$ und $A > 0$ where $\phi_{m,n}(A+\delta) \ll a_{m,n} \ll \phi_{m,n}(A-\delta)$ for every $\delta \in (0,A)$. Hence, we have

$$\lim_{n\to\infty} \sum_{m\in \mathbb{M}_{k,n}} a_{m,n} \frac{1}{m} = \frac{q}{A} .$$

for all $k \in \mathbb{N}_0$.

Proof. Since the limit q exist and q is positiv, for every $\epsilon \in (0,1)$ there is an index $n_0 \in \mathbb{N}$ such that

$$0 < (1-\epsilon)q \leq \frac{|\mathbb{M}_{0,k}|}{k} \leq (1+\epsilon)q$$

for all $k \geq n_0$. Thus, we obtain

$$\frac{1-\epsilon}{1+\epsilon} \frac{m}{n} \leq \frac{|\mathbb{M}_{0,m}|}{|\mathbb{M}_{0,n}|} \leq \frac{1+\epsilon}{1-\epsilon} \frac{m}{n} ,$$

for all $n_0 \leq m \leq n$. Thus, assertion a) holds true.

Choose arbitrary values of $b>0$. According to assertions a) and d) of Lemma A.5 and Lemma A.2 c3) we obtain

$$\lim_{n\to\infty} \sum_{m\in \mathbb{M}_{k,n}} \phi_{m,n}(b) \cdot \frac{1}{m}$$

$$= \lim_{n\to\infty} \sum_{m\in \mathbb{M}_{k,n}} (\frac{|\mathbb{M}_{0,m}|}{|\mathbb{M}_{0,n}|})^b \cdot \frac{1}{|\mathbb{M}_{0,m}|} \cdot \frac{|\mathbb{M}_{0,m}|}{m}$$

$$= \lim_{n\to\infty} \sum_{i=|\mathbb{M}_{0,k}|+1}^{|\mathbb{M}_{0,n}|} (\frac{i}{|\mathbb{M}_{0,n}|})^b \frac{1}{i} \cdot q = \frac{q}{b} .$$

for all $k\in \mathbb{N}_0$.

Due to the assertion concerning $(a_{m,n})_{m\leq n}$ and because of Lemma A.2 c1), c2) we have

$$\frac{q}{A+\delta} \leq \lim_{n\to\infty} \sum_{m\in \mathbb{M}_{k,n}} a_{m,n} \frac{1}{m} \leq \frac{q}{A-\delta}$$

for all $\delta \in (0,A)$ and $k\in \mathbb{N}_0$.

Let $(A_n)_n$ be a sequence of $\mathcal{P}$ such that the limit

$$A^{(k)} := \lim_{\mathbb{N}^{(k)} \ni n\to\infty} A_n$$

exist for every $k \in \{1,\dots,\kappa\}$ and this limit is positive. We define

$$b_{m,n} := \begin{cases} \prod\limits_{j=m+1}^{n} (1 - \dfrac{A_j}{j}), & \text{for } m > n \\[2em] 1 & \text{, for } m=n. \end{cases}$$

On this double sequence $(b_{m,n})_{m \le n}$ we have the following result:

<u>Lemma A.7.</u> Let $t \in \mathbb{R}$ be such that $A := q_1 A^{(1)} + \ldots + q_\kappa A^{(\kappa)} - t$ is positive. Hence,

a) $\phi_{m,n}(A+\delta) \ll b_{m,n} \cdot (\dfrac{n}{m})^t \ll \phi_{m,n}(A-\delta)$ for all $\delta \in (0,A)$,

b) $\lim\limits_{n \to \infty} b_{m,n} \cdot n^t = 0$ for all $m \in \mathbb{N}_o$,

c) $\lim\limits_{n \to \infty} \sum\limits_{m \in \mathbb{M}_{k,n}} b_{m,n} \cdot (\dfrac{n}{m})^t \dfrac{1}{m} = \dfrac{q}{A}$ for all $k \in \mathbb{N}_o$.

<u>Proof.</u> If $b_1, \ldots, b_\kappa$ are arbitrarily chosen positive numbers, then Lemma A.2 b), Lemma A.5 a), and Lemma A.6 a) yield for all $m \le n$

$$\prod_{k=1}^{\kappa} \prod_{\ell \in \mathbb{N}_{m,n}^{(k)}} (1 - \frac{b_k}{|\mathbb{N}_{o,\ell}^{(k)}|}) =$$

$$= \prod_{k=1}^{\kappa} \prod_{i=|\mathbb{N}_{o,m}^{(k)}|+1}^{|\mathbb{N}_{o,n}^{(k)}|} (1 - \frac{b_k}{i}) \sim \prod_{k=1}^{\kappa} (\frac{|\mathbb{N}_{o,m}^{(k)}|}{|\mathbb{N}_{o,n}^{(k)}|})^{b_k}$$

$$\sim \prod_{k=1}^{\kappa} (\frac{m}{n})^{b_k} = (\frac{m}{n})^{b_1 + \ldots + b_\kappa} . \tag{*}$$

As a result of the equation

$$b_{m,n} = \prod_{k=1}^{\kappa} \left(\prod_{\ell \in \mathbb{N}_{m,n}^{(k)}} (1 - \frac{A_\ell \cdot \frac{|\mathbb{N}_{o,\ell}^{(k)}|}{\ell}}{|\mathbb{N}_{o,\ell}^{(k)}|}) \right)$$

and due to the existence of an index $n_o \in \mathbb{N}$ for a given sufficiently small value of $\delta > 0$ such that

$$0 < q_k A^{(k)} - \frac{\delta}{\kappa} \le A_\ell \cdot \frac{|\mathbb{N}_{o,\ell}^{(k)}|}{\ell} \le q_k \cdot A^{(k)} + \frac{\delta}{\kappa}$$

for all $\ell \ge n_o$, $\ell \in \mathbb{N}^{(k)}$ and $k \in \{1, \ldots, \kappa\}$, we find - according to (*) -

$$(\tfrac{m}{n})^{A+\delta} \ll b_{m,n} \cdot (\tfrac{n}{m})^{t} \ll (\tfrac{m}{n})^{A-\delta} \qquad (**)$$

for every $\delta \in (0,A)$ from which follows assertion a) according to Lemma A.5 a).

(**) confirms the validity of assertion b), and c) is valid because of a) and Lemma A.6 b).

We have to consider one more proposition on another double sequence:

<u>Lemma A.8.</u> Assume that $a, b \in \mathbb{R}$, $x \in \mathbb{R}$ with $|x| < 1$ and $(a_{m,n})_{m \leq n}$ is a real double sequence such that

a) $(\tfrac{m}{n})^{a} \ll a_{m,n} \ll (\tfrac{m}{n})^{b}$

b) $\lim\limits_{n \to \infty} x^{n} \cdot a_{m,n} = 0$ for all $m \in \mathbb{N}$.

Hence, for all $k \in \mathbb{N}$

$$\lim_{n \to \infty} \sum_{m=k}^{n} x^{n-m} \cdot a_{m,n} = \frac{1}{1-x} \; .$$

<u>Proof.</u> By separate summation of odd and even indices of the above-mentioned sum, the proof may be reduced to the case $x \geq 0$. Due to assumptions a) and b) we only have to show that

$$\lim_{n \to \infty} \sum_{m=k}^{n} x^{n-m} (\tfrac{m}{n})^{c} = \frac{1}{1-x} \; , \qquad (*)$$

for all $k \in \mathbb{N}$ and $c \in \mathbb{R}$, where $x \in [0,1)$. For given $x \in [0,1)$, $k \in \mathbb{N}$ we consider for $n \geq k$ the function

$$\varphi_{n}(c) := \sum_{m=k}^{n} x^{n-m} (\tfrac{m}{n})^{c} \; .$$

A simple computation process yields

$$\lim_{n \to \infty} \varphi_{n}(0) = \frac{1}{1-x} \text{ and} \qquad (i)$$

$$\varphi_{n}(c) = \varphi_{n}(c+1) + \frac{1}{n} \cdot R_{n}(c), \qquad (ii)$$

where

$$R_{n}(c) := \sum_{m=k}^{n-1} (n-m) x^{n-m} (\tfrac{m}{n})^{c}.$$

Because of the inequalities

$$|R_n(c)| \leq \sum_{m=k}^{n-1} (n-m) \cdot x^{n-m} (\frac{m}{n})^{-|c|} \leq \sum_{m=k}^{n-1} (n-m)(n-m+1)^{|c|} \cdot x^{n-m} -$$

$$- \sum_{i=1}^{n-k} i(i+1)^{|c|} x^i \leq \sum_{i=1}^{\infty} i(i+1)^{|c|} x^i < \infty,$$

and because of (ii) we obtain

$$\lim_{n \to \infty} \varphi_n(c) - \lim_{n \to \infty} \varphi_n(c+1), \qquad (iii)$$

from which follows according to (i)

$$\lim_{n \to \infty} \varphi_n(1) - \frac{1}{1-x} . \qquad (iv)$$

According to (iii) there is a number $\bar{c} \in [0,1]$ such that

$$\lim_{n \to \infty} \varphi_n(c) - \lim_{n \to \infty} \varphi_n(\bar{c}). \qquad (v)$$

Since $\bar{c} \in [0,1]$, we obtain

$$(\frac{m}{n})^1 \leq (\frac{m}{n})^{\bar{c}} \leq (\frac{m}{n})^o - 1,$$

for all $m \leq n$. Thus, (i) and (iv) yield

$$\lim_{n \to \infty} \varphi_n(\bar{c}) - \frac{1}{1-x} ,$$

Equation (*) is thus verified by (v).

Annex B. A central limit theorem.

A limit theorem on a sequence of dependent random vectors is required in

section 3. The theorem provided in this paper is a ν-dimenional version

of Brown's limit theorem on martingale difference schemata (cf. [5],

Theorem 9.2.3).

Let $(\Omega, \mathcal{O\!l}, P)$ be a P-space and $(\mathcal{O\!l}_n)_{n\in \mathbb{N}}$ be an isotone sequence of σ-al-

gebras of $\mathcal{O\!l}$, i.e. we have $\mathcal{O\!l}_1 \subseteq \mathcal{O\!l}_2 \subseteq \ldots \subseteq \mathcal{O\!l}_n \subseteq \mathcal{O\!l}$ for all $n\in \mathbb{N}$.

We further assume that $(Z_n)_{n\in \mathbb{N}}$ is a sequence of mappings of Ω into $\mathbb{R}^\nu$,

such that Z_n is $\mathcal{O\!l}_{n+1}$-measurable for all $n\in \mathbb{N}$, and we find

$\quad E(Z_n|\mathcal{O\!l}_n) \equiv 0.$ \hfill (B1)

Let $(A_{m,n})_{m\leq n}$ be a given sequence of real symmetric $\nu\times\nu$ matrices from

which we obtain now the following sequence of random vectors:

$(\sum_{m=1}^{n} A_{m,n} \cdot Z_m)_{n\in \mathbb{N}}$

For this sequence the following central limit theorem holds true:

Theorem B.1. If

$\quad$(kV): $\sum_{m=1}^{n} A_{m,n} \cdot E(Z_m Z_m^T|\mathcal{O\!l}_m) \cdot A_{m,n}$ converges stochastically to a

symmetric deterministic $\nu\times\nu$ matrix V,

$\quad$(kL): $\sum_{m=1}^{n} ||A_{m,n}||^2 E(||Z_m||^2 \cdot I_{(||A_{m,n} Z_m||>\epsilon)}|\mathcal{O\!l}_m)$

converges stochastically to 0 for all $\epsilon>0$, where for $B \subseteq \Omega$

$\quad I_B(\omega) := \{ \begin{matrix} 1, & \text{for } \omega\in B \\ 0, & \text{otherwise} \end{matrix}$

is the characteristic function of B, then the sequence $(\sum_{m=1}^{n} A_{m,n} \cdot Z_m)_{n\geq 1}$

converges - in distribution - to an N(0,V)-distributed random vector.

Proof. We assume that D is an arbitrary N(o,V)-distributed random vector

and $0\neq\lambda\in \mathbb{R}^\nu$ is an arbitrary fixed vector. We take

$$Y_{m,n} := \lambda^T \cdot A_{m,n} \cdot Z_m, \quad Y_n := \lambda^T \sum_{m=1}^{n} A_{m,n} \cdot Z_m \text{ and}$$

$$Y := \lambda^T \cdot D \text{ for } m, n \in \mathbb{N} \text{ and } m \le n.$$

According to the "Cramér-Wold-device" (cf.[5], Theorem 8.7.6) we only have to show that $(Y_n)_n$ - in distribution - converges to Y. Because of (B1), $(Y_{m,n}, \mathcal{O}_{m+1})$ is a martingale difference scheme, therefore - according to Brown's theorem -

$$Y_n = \sum_{m=1}^{n} Y_{m,n} \text{ converges in distribution to an } N(o, \lambda^T \cdot V \cdot \lambda)\text{-distributed}$$

random variable Y, if

$(kV)_1: \sum_{m=1}^{n} E(Y_{m,n}^2 | \mathcal{O}_m)$ converges stochastically to $\lambda^T \cdot V \cdot \lambda$, and

$(kL)_1: \sum_{m=1}^{n} E(Y_{m,n}^2 \cdot I_{(|Y_{m,n}| > \epsilon')} | \mathcal{O}_m)$ converges stochastically to 0

for all $\epsilon' > 0$.

$(kV)_1$ and $(kL)_1$ follow, however, from (kV) and (kL), since

$$E(Y_{m,n}^2 | \mathcal{O}_m) = \lambda^T A_{m,n} E(Z_m Z_m^T | \mathcal{O}_m) A_{m,n} \lambda$$

and

$$E(Y_{m,n}^2 \cdot I_{(|Y_{m,n}| > \epsilon')} | \mathcal{O}_m) \le$$

$$\le ||\lambda||^2 \, ||A_{m,n}||^2 \, E(||Z_m||^2 \cdot I_{(||A_{m,n} Z_m|| > \frac{\epsilon'}{||\lambda||})} | \mathcal{O}_m)$$

for all $m \le n$ and $\epsilon' > 0$.

Sufficient conditions applying to (kV) and (kL) in Theorem B.1 are provided by the next theorem.

Choose a particular sequence of matrices $(A_{m,n})_{m \le n}$: We assume a given decomposition $\mathbb{N} = \mathbb{N}^{(1)} \cup \ldots \cup \mathbb{N}^{(\kappa)}$ of $\mathbb{N}$ into disjoint non-finite subsets and given symmetric matrices $W^{(1)}, \ldots, W^{(\kappa)}$. The sequence $(A_{m,n})_{m \le n}$ may now be described

a) $\lim_{n \to \infty} ||A_{m,n}|| = 0$ for all $m \in \mathbb{N}$,   (B2.1)

b) there is an index n_o and there are positive sequences $(\alpha_m)_{n_o \le m}$ and

   $(a_{m,n})_{n_o \le m \le n}$ such that

$$||A_{m,n}||^2 \leq a_{m,n}^2 \text{ for all } n_o \leq m \leq n, \qquad (B2.2)$$

$$\lim_{n\to\infty} \sum_{m=n_o}^{n} a_{m,n}^2 < \infty, \qquad (B2.3)$$

$$||A_{m,n}|| \leq \alpha_m \text{ for all } n_o \leq m \leq n, \qquad (B2.4)$$

c) there is a limit matrix

$$V: = \lim_{n\to\infty} \sum_{m=1}^{n} A_{m,n} \cdot W_m \cdot A_{m,n}, \qquad (B2.5)$$

where $W_m: = W^{(k)}$ for $m \in \mathbb{N}^{(k)}$ with $k \in \{1, \ldots, \kappa\}$.

<u>Theorem B.2.</u> We assume that $(A_{m,n})_{m \leq n}$ satisfies the premises (B2).

a) If

$$\lim_{m\to\infty} E(||Z_m||^2 \cdot I_{(||Z_m|| > \frac{\epsilon}{\alpha_m})}) = 0 \qquad (B3)$$

or

$$\lim_{m\to\infty} E(||Z_m||^2 \cdot I_{(||Z_m|| > \frac{\epsilon}{\alpha_m})} | \mathcal{O}_m) = 0 \text{ almost sure} \qquad (B4)$$

for every $\epsilon > 0$, then (kL) is fulfilled.

b) If

$$\lim_{\mathbb{N}^{(k)} \ni m\to\infty} E(Z_m Z_m^T) = W^{(k)}, \qquad (B5.1)$$

$$\lim_{m\to\infty} E||E(Z_m Z_m^T|\mathcal{O}_m) - E Z_m Z_m^T|| = 0 \qquad (B5.2)$$

or

$$\lim_{\mathbb{N}^{(k)} \ni m\to\infty} E(Z_m Z_m^T|\mathcal{O}_m) = W^{(k)} \text{ almost sure} \qquad (B6)$$

for $k = 1, \ldots, \kappa$, then (kV) is fulfilled.

<u>Proof.</u> Because of (B2.4) we find

$$||Z_m||^2 \cdot I_{(||A_{m,n} Z_m|| > \epsilon)} \leq ||Z_m||^2 \cdot I_{(||Z_m|| > \frac{\epsilon}{\alpha_m})},$$

for $n_o \leq m \leq n$, $\epsilon > 0$. If (B3) is fulfilled, we obtain according to (B2.1), (B2.2),(B2.3), and Lemma A2c)

$$\lim_{n\to\infty} E U_n = 0,$$

or, if (B4) is fulfilled, then

$$\lim_{n \to \infty} U_n = 0 \quad \text{almost sure}$$

where

$$U_n := \sum_{m=1}^{n} ||A_{m,n}||^2 \cdot E(||Z_m||^2 \cdot I_{(||A_{m,n} Z_m||>\epsilon)} |\alpha_m) .$$

For the random matrix

$$V_n := \sum_{m=1}^{n} A_{m,n} \cdot E(Z_m Z_m^T |\alpha_m) \cdot A_{m,n}$$

because of (B2) condition (B6) yields

$$\lim_{n \to \infty} ||V_n - V|| = 0 \quad \text{almost sure,} \tag{i}$$

and condition (B5) yields

$$\lim_{n \to \infty} ||EV_n - V|| = 0,$$

$$\lim_{n \to \infty} E||V_n - EV_n|| = 0.$$

Hence, we obtain - according to the triangular inequality -

$$\lim_{n \to \infty} E||V_n - V|| = 0. \tag{ii}$$

(kV) finally follows from (i) or (ii).

References

[1] Bauer H.: Wahrscheinlichkeitstheorie und Grundzüge der Maßtheorie.
Walter de Gruyter, Berlin 1974

[2] Blum J.R.: Multidimensional Stochastic Approximation Methods.
Ann.Math.Stat. 25 (1954) 737-744

[3] Chung K.L.: On a Stochastic Approximation Method.
Ann.Math.Stat. 25 (1954) 463-483

[4] Fabian V.: On Asymptotic Normality in Stochastic Approximation.
Ann.Math.Stat. 39 (1968) 1327-1332

[5] Gänssler P., Stute W.: Wahrscheinlichkeitstheorie.
Springer-Verlag, Berlin 1977

[6] Knopp K.: Theorie und Anwendung der unendlichen Reihen.
Springer-Verlag, Berlin 1964

[7] Marti K.: Approximationen stochastischer Optimierungsprobleme.
Verlag A. Hain, Königstein / Ts. 1979

[8] Marti K., Fuchs E.: Rates of Convergence of Semi-Stochastic Approximation Procedures for Solving Stochastic Optimization Problems. Optimization 17 (1986) 2, 243-265

[9] Marti K.: Stochastische Optimierung II. Vorlesung an der Universität der Bundeswehr München, Neubiberg, Wintertrimester 1987

[10] Marti K., Plöchinger E.: Optimal Step Sizes in Semi-Stochastic Approximation Procedures. Forschungsschwerpunkt Simulation und Optimierung deterministischer und stochastischer dynamischer Systeme, Universität der Bundeswehr München, Neubiberg 1988

[11] Pflug G.: Stochastic Minimization with constant Step-Size.
SIAM Journal of Control 24 (1986), 655-666

[12] Robbins H., Monro S.: A Stochastic Approximation Method.
Ann.Math.Stat. 22 (1951), 400-407

[13] Sacks J.: Asymptotic Distribution of Stochastic Approximation
Procedures. Ann.Math.Stat. 29 (1958) 373-405

[14] Wasan M.T.: Stochastic Approximation.
Cambridge University Press 1969

CONTINUITY AND STABILITY IN TWO-STAGE STOCHASTIC INTEGER PROGRAMMING

Rüdiger Schultz
Humboldt-Universität zu Berlin
Fachbereich Mathematik
PSF 1297, D–1086 Berlin

ABSTRACT: For two-stage stochastic programs where the optimization problem in the second stage is a mixed–integer linear program continuity of the expectation of second–stage costs jointly in the first–stage strategy and the integrating probability measure is derived. Then, regarding the two–stage stochastic program as a parametric program with the underlying probability measure as parameter, continuity of the locally optimal value and upper semicontinuity of the corresponding set of local solutions are established.

1 Introduction

In this paper, we will analyse parameter dependent two–stage stochastic optimization problems of the type

$$P(\mu) \qquad \min\{f(x) + Q(x,\mu) : x \in C\},$$

where

$$Q(x,\mu) = \int\limits_{\mathbf{R}^s} \Phi(z - Ax)\mu(dz) \qquad (1.1)$$

and

$$\Phi(b) = \min\{q^T y + q'^T y' \ : \ Wy + W'y' = b, \ y' \geq 0, \ y \geq 0, \ y \in \mathbf{Z}^{\tilde{s}}\} \qquad (1.2)$$

Here we assume that f is a continuous real–valued function on $\mathbf{R}^m$, $C \subset \mathbf{R}^m$ non–empty, closed, $z \in \mathbf{R}^s$, $A \in L(\mathbf{R}^m, \mathbf{R}^s)$, $q \in \mathbf{R}^{\tilde{s}}$, $q' \in \mathbf{R}^{s'}$, $W \in L(\mathbf{R}^{\tilde{s}}, \mathbf{R}^s)$, $W' \in L(\mathbf{R}^{s'}, \mathbf{R}^s)$, $b \in \mathbf{R}^s$. By $\mathbf{Z}^{\tilde{s}}$ we denote the subset of vectors in $\mathbf{R}^{\tilde{s}}$ having only integral components. Throughout, we assume that W and W' are rational matrices. The underlying measure μ, which is supposed to belong to the set $\mathcal{P}(\mathbf{R}^s)$ of all (Borel) probability measures on $\mathbf{R}^s$, enters the above model as a parameter. Further assumptions to make (1.1) and (1.2) well–defined are given below.

The essential difference between $P(\mu)$ and conventional stochastic programs with linear recourse (cf. e. g. [5], [6], [14]) lies in the integrality constraints of the second–stage (or recourse) program (1.2). Understanding the second stage as an optimization process where deviations between predictions and realizations of random data may call for integral corrective actions one arrives at a model $P(\mu)$.

In the literature, the discussion of two-stage stochastic integer programs like $P(\mu)$ started in [9], [13], where principal difficulties stemming from the integrality constraints were emphasized. Such difficulties include that, for fixed $\mu \in \mathcal{P}(\mathbf{R}^s)$, the function $Q(.,\mu)$ in (1.1) is neither convex nor continuous in general. Without integrality constraints the integrand Φ (cf. (1.2)) is a positively homogeneous convex function, with such constraints the value function need not be continuous ([2], [4]). However, analysing this value function in more detail, it can be shown that $Q(.,\mu)$ is continuous on $\mathbf{R}^m$ provided that $P(\mu)$ is well–defined and μ is absolutely continuous with respect to the Lebesgue measure on $\mathbf{R}^s$ (cf. Theorem 3.2 in [12] and, for a first result along this line, Theorem 5.1 in [13]). Ensuring $Q(.,\mu)$ to be locally Lipschitzian on $\mathbf{R}^m$ requires additional hypotheses on μ, which are introduced and justified in [12] (Theorem 3.5, Examples 3.7, 3.8).

In the present paper, we analyse continuity of the function Q jointly in x and μ, when equipping $\mathcal{P}(\mathbf{R}^s)$ with the topology of weak convergence of probability measures (cf. [3]). Then, we present consequences of this joint continuity for the stability of the problem $P(\mu)$, i. e. we derive continuity properties of the mappings assigning to $\mu \in \mathcal{P}(\mathbf{R}^s)$ local optimal values and sets of local minimizers of $P(\mu)$. As for stochastic programs without integrality constraints, such a stability analysis is mainly motivated twofold. Namely, when formulating the model one often has incomplete information about the underlying probability measure and when designing solution procedures a possible approach is to approximate "complicated" measures by "simpler" ones. In both cases one wants to be sure that "small" changes in the measure cause only "small" changes in the optimum and the optimizers, respectively. The general stability theory for minimization problems, which we are going to employ, also covers programs whose perturbations have only lower semicontinuous objectives. In the context of two–stage stochastic integer programs this is essential, since perturbations are often connected with discrete measures, and, as we will see below, a discrete measure μ leads to a lower semicontinuous function $Q(.,\mu)$ in (1.1). Let us further remark that the counterpart to our analysis for two–stage stochastic programs without integrality constraints has been elaborated in [7] and [11]. The main difference between these contributions and the present one is that the joint continuity of the counterpart to our functional Q (cf. (1.1)) is established in a way which, due to the discontinuity of the integrand Φ (cf. (1.2)), cannot be followed here.

2 Continuity

Before starting the continuity analysis of Q it is necessary to collect a few prerequisites about the integrand Φ in (1.2) being a value function of a parametric linear mixed-integer program. As basic references in this respect we refer to the monograph [2] and to the article [4].

First we assume that for each $b \in \mathbf{R}^s$ the constraint set of the optimization problem defining $\Phi(b)$ is nonempty. Let further $\Phi(0) = 0$. Then, $\Phi(b) \in \mathbf{R}$ for all $b \in \mathbf{R}^s$. Moreover, it holds (with $\|.\|$ denoting the Euclidean norm):

Proposition 2.1 *([2] Th. 8.1, [4] Th. 2.1)*
There exist constants $\alpha > 0$, $\beta > 0$ such that for all $b', b'' \in \mathbf{R}^s$ we have

$$\|\Phi(b') - \Phi(b'')\| \le \alpha \|b' - b''\| + \beta.$$

Proposition 2.2 *([4] Th. 3.3)*
There exist constants $\tau > 0$, $\delta > 0$ and vectors $d_1, \ldots, d_l \in \mathbf{R}^s$, $\tilde{d}_1, \ldots, \tilde{d}_{l'} \in \mathbf{R}^s$ such that for all $b \in \mathbf{R}^s$

$$\Phi(b) = \min_y \{q^T y + \max_{j \in \{1, \ldots, l\}} d_j^T(b - Wy) \; : \; y \in Y(b)\},$$

where

$$Y(b) = \{y \in \mathbf{Z}^s \; : \; y \ge 0, \; \Sigma|y_i| \le \tau\Sigma|b_r| + \delta, \; \tilde{d}_k^T(b - Wy) \ge 0, \; k = 1, \ldots, l'\}.$$

If there exist $\bar{b} \in \mathbf{R}^s$ and an open neighbourhood of $\bar{b}$ on which $Y(.)$ remains constant, then Proposition 2.2 says that, on the mentioned neighbourhood, Φ is the pointwise minimum of finitely many continuous piecewise linear functions and, hence, continuous at $\bar{b}$. If $b \in \mathbf{R}^s$ is such that $Y(.)$ does not remain constant on any open neighbourhood of b, then there must exist $y \in \mathbf{Z}^s$, $y \ge 0$ such that at least one of the inequalities

$$\Sigma|y_i| \le \tau\Sigma|b_r| + \delta$$

and

$$\tilde{d}_k^T(b - Wy) \ge 0, \quad k = 1, \ldots, l'$$

holds as an equation.

Therefore, we conclude that the set of discontinuity points of Φ is contained in a countable union of hyperplanes in $\mathbf{R}^s$.

Now let us make precise our assumptions to have (1.1) and (1.2) well–defined. We suppose:

There exists a $u \in \mathbf{R}^s$ such that $W^T u \leq q$ and $W'^T u \leq q'$. $\qquad (2.1)$

For all $t \in \mathbf{R}^s$ there exist $y \in \mathbf{Z}^{\bar{s}}, y' \in \mathbf{R}^{s'}$ such that
$y \geq 0,\ y' \geq 0$ and $Wy + W'y' = t$. $\qquad (2.2)$

It holds that $\displaystyle\int_{\mathbf{R}^s} \|z\| \mu(dz) < +\infty.$ $\qquad (2.3)$

Observe that these assumptions are quite similar to those usually imposed for linear stochastic programs with complete recourse when not taking into account integrality constraints. Indeed, (2.1) corresponds to "dual feasibility", (2.2) guarantees "primal feasibility" and, due to the integrability assumption (2.3), the integral in (1.1) should be finite.

However, a correct reasoning using results from linear mixed–integer programming must be given: The Assumptions (2.1) and (2.2) together with the duality theorem of linear programming and Lemma 7.1 in [2] imply that $\Phi(z - Ax) \in \mathbf{R}$ for all $z \in \mathbf{R}^s$, $x \in \mathbf{R}^m$. As a consequence of (2.3) and Proposition 2.1 we obtain that $Q(.,\mu)$ is a real–valued function on $\mathbf{R}^m$ (cf. Lemma 3.1 in [12] for the detailed argument). Finally, we note that (2.1) is equivalent to $\Phi(0) = 0$.

To analyse the joint continuity of Q in x and μ, we first have to explain what kind of convergence is considered on $\mathcal{P}(\mathbf{R}^s)$. A convenient notion in this respect is that of weak convergence of probability measures. A sequence $\{\mu_n\}$ of probability measures in $\mathcal{P}(\mathbf{R}^s)$ is said to converge weakly to $\mu \in \mathcal{P}(\mathbf{R}^s)$, i. e. $\mu_n \xrightarrow{w} \mu$, if for any bounded continuous function $g : \mathbf{R}^s \longrightarrow \mathbf{R}$ we have

$$\int_{\mathbf{R}^s} g(z)\mu_n(dz) \longrightarrow \int_{\mathbf{R}^s} g(z)\mu(dz) \quad \text{as} \quad n \to \infty.$$

For a detailed description of the topology of weak convergence of probability measures we refer to the monograph [3].

Before formulating the continuity result for Q we introduce, for notational convenience, the subset of probability measures

$$\Delta_{p,K}(\mathbf{R}^s) := \{\mu' \in \mathcal{P}(\mathbf{R}^s) : \int_{\mathbf{R}^s} \|z\|^p \mu'(dz) \leq K\},$$

where $p > 1$ and $K > 0$ are fixed real numbers.

Theorem 2.3 *Assume (2.1), (2.2) and let $\mu \in \Delta_{p,K}(\mathbf{R}^s)$ for some $p > 0$, $K > 0$. If μ is absolutely continuous with respect to the Lebesgue measure, then Q, as a function from $\mathbf{R}^m \times \Delta_{p,K}(\mathbf{R}^s)$ to $\mathbf{R}$, is continuous on $\mathbf{R}^m \times \{\mu\}$, where convergence on $\Delta_{p,K}(\mathbf{R}^s)$ is understood as weak convergence.*

Proof: Let $x \in \mathbf{R}^m$ and consider sequences $\{x_n\}$, $\{\mu_n\}$ in $\mathbf{R}^m$ and $\Delta_{p,K}(\mathbf{R}^s)$, respectively, such that $x_n \to x$ and $\mu_n \overset{w}{\to} \mu$ as $n \to \infty$. We introduce functions $h_n : \mathbf{R}^s \to \mathbf{R}$ and $h : \mathbf{R}^s \to \mathbf{R}$ defined by

$$h_n(z) := \Phi(z - Ax_n) \qquad \text{and} \qquad h(z) := \Phi(z - Ax).$$

Since the value function Φ is lower semicontinuous ([4] p. 133), the functions h_n and h are measurable.

Let E denote the set of all those $z \in \mathbf{R}^s$ such that there exists a sequence $\{z_n\}$ in $\mathbf{R}^s$ with

$$z_n \to z \qquad \text{and} \qquad h_n(z_n) \nrightarrow h(z).$$

If Φ is continuous at $z - Ax$, then, obviously, $z \notin E$. Hence, E is contained in the set of all those $z \in \mathbf{R}^s$ such that Φ is not continuous at $z - Ax$. In view of Proposition 2.2, cf. the argument given there, the latter set is contained in a countable union of hyperplanes in $\mathbf{R}^s$. Since μ is absolutely continuous with respect to the Lebesgue measure and E is measurable (for a proof see [3] p. 226), we obtain $\mu(E) = 0$.

The functions h_n and h now induce measures $\mu_n h_n^{-1}$ and μh^{-1} on $\mathbf{R}$ which are given by

$$(\mu_n h_n^{-1})(B) := \mu_n(h_n^{-1}(B)) \quad \text{and} \quad (\mu h^{-1})(B) := \mu(h^{-1}(B)),$$

where B is an arbitrary Borel set in $\mathbf{R}$.

By Rubin's Theorem ([3] Th. 5.5, p. 34), weak convergence $\mu_n \overset{w}{\to} \mu$ and $\mu(E) = 0$ imply that

$$\mu_n h_n^{-1} \overset{w}{\longrightarrow} \mu h^{-1}.$$

Consider some probability space $(\Omega, \mathcal{B}, P)$ and the random variables $X_n, X : (\Omega, \mathcal{B}, P) \to \mathbf{R}$ having distributions $\mu_n h_n^{-1}$ and μh^{-1}. We show that

$$\lim_{a \to \infty} \sup_n \int_{\{\omega : |X_n(\omega)| \geq a\}} |X_n(\omega)| P(d\omega) = 0. \tag{2.4}$$

For any $a \geq 0$ we have

$$\int\limits_{\Omega} |X_n|^p dP \geq \int\limits_{\{|X_n|\geq a\}} |X_n| \cdot |X_n|^{p-1} dP \geq a^{p-1} \int\limits_{\{|X_n|\geq a\}} |X_n| dP$$

and, hence,

$$\int\limits_{\{|X_n|\geq a\}} |X_n| dP \leq a^{1-p} \int\limits_{\Omega} |X_n|^p dP. \tag{2.5}$$

Furthermore,

$$\int\limits_{\Omega} |X_n|^p dP = \int\limits_{\mathbf{R}} |z'|^p \mu_n h_n^{-1}(dz'),$$

and by change of variable the latter expression equals

$$\int\limits_{\mathbf{R}^s} |h_n(z)|^p \mu_n(dz).$$

Recalling that $\Phi(0) = 0$ and using Proposition 2.1 we obtain

$$\begin{aligned}
|h_n(z)|^p &= |\Phi(z - Ax_n) - \Phi(0)|^p \\
&\leq (\alpha \|z - Ax_n\| + \beta)^p \\
&\leq (\alpha \|z\| + \alpha \|Ax_n\| + \beta)^p.
\end{aligned}$$

Since the set $\{\|Ax_n\| : n = 1, 2, \ldots\}$ is bounded and all μ_n belong to $\Delta_{p,K}(\mathbf{R}^s)$, we have a positive constant c such that

$$\int\limits_{\mathbf{R}^s} |h_n(z)|^p \mu_n(dz) \leq c \quad \text{for all} \quad n = 1, 2, \ldots.$$

Together with (2.5) this yields (2.4).

Now, we are in the position to apply Theorem 5.4 in [3] (p. 32). The theorem says that $\mu_n h_n^{-1} \xrightarrow{w} \mu h^{-1}$ and (2.4) together provide

$$\int\limits_{\Omega} X_n dP \longrightarrow \int\limits_{\Omega} X dP \qquad \text{as} \qquad n \to \infty. \tag{2.6}$$

Rephrasing the integrals and changing variables we obtain

$$\int\limits_{\Omega} X_n dP = \int\limits_{\mathbf{R}} z' \mu_n h_n^{-1}(dz') = \int\limits_{\mathbf{R}^s} h_n(z) \mu_n(dz)$$

and

$$\int\limits_{\Omega} X dP = \int\limits_{\mathbf{R}^s} h(z) \mu(dz).$$

Hence, (2.6) means

$$Q(x_n, \mu_n) \longrightarrow Q(x, \mu) \qquad \text{as} \qquad n \to \infty.$$

q.e.d.

Remark 2.4 *The assumption in Theorem 2.3 that μ be absolutely continuous is indispensable, since e. g. a discrete measure μ with finitely many mass points will lead to a function $Q(., \mu)$ that is in general a finite convex combination of discontinuous functions and, hence, again discontinuous.*

Remark 2.5 *Without any assumption that finally leads to some uniform integrability as in (2.4) we cannot expect to end up with the desired continuity of Q. This can already be observed in a corresponding analysis for continuous integrands (cf. e. g. [11], p. 1410).*

Remark 2.6 *Of course, the above theorem remains valid when $\bar{s} = 0$, i.e. without integrality constraints in the second stage. In this respect, it contains corresponding continuity results in [7] and [11] as special cases, and it shows simultaneously how to relax the continuity of the integrand Φ, which is a basic assumption in [7], [11].*

3 Stability

Since, for a fixed measure μ, the function $Q(., \mu)$ is in general non-convex, our stability considerations must include the analysis of local solutions.
The problem

$$P(\mu) \qquad \min\{f(x) + Q(x, \mu) : x \in C\}$$

is now understood as a parametric optimization problem with the parameter μ varying in $\mathcal{P}(\mathbf{R}^s)$ or some subset.
To study stability of local solutions we introduce the following localized versions of the optimal–value function and the solution set mapping

$$\varphi_V(\mu) := \inf\{f(x) + Q(x, \mu) : x \in C \cap \mathrm{cl}\, V\},$$
$$\psi_V(\mu) := \{x \in C \cap \mathrm{cl}\, V : f(x) + Q(x, \mu) = \varphi_V(\mu)\}.$$

The set V arising above is an arbitrary subset of $\mathbf{R}^m$ and $\mathrm{cl}\, V$ denotes the closure of V.

As pointed out in [8], [10], it is crucial for the stability analysis of local solutions to an optimization problem that the considerations include all local minimizers that are, in some sense, near the minimizers one is interested in. This leads to the concept of a complete local minimizing set (CLM set), which can be formulated in our terminology as follows: Given $\mu \in \mathcal{P}(\mathbf{R}^s)$, a non–empty set $M \subset \mathbf{R}^m$ is called a CLM set for $P(\mu)$ with respect to an open set $V \subset \mathbf{R}^m$ if $M \subset V$ and $M = \psi_V(\mu)$. Examples of CLM sets are the set of global minimizers and strict local minimizers. For more details consult [10], [8].

As we will see below, local minimizers behave stable under perturbations if they form a bounded CLM set for the unperturbed problem. For general parametric programs there exist examples showing that neither boundedness nor the CLM property can be omitted without losing stability. Due to the following example, also in the setting of $P(\mu)$, stability is not guaranteed for bounded local solution sets without CLM property.

Let

$$P(\mu) \qquad \min\{4x + Q(x, \mu) : x \in \mathbf{R}\},$$

where

$$Q(x, \mu) = \int_{\mathbf{R}} \Phi(z + x)\mu(dz),$$

$$\Phi(b) = \min\{-y : y \le b, y \in \mathbf{Z}\}$$

and μ is the uniform distribution on $[0, \frac{1}{4}]$.
One calculates

$$Q(x, \mu) = \begin{cases} 0 & 0 \le x \le \frac{3}{4} \\ -4x + 3 & \frac{3}{4} \le x \le 1 \\ -1 & 1 \le x \le \frac{7}{4}. \end{cases}$$

Hence, $x = \frac{7}{8}$ is a local minimizer to $P(\mu)$, but, of course, there is no open set $V \subset \mathbf{R}$ such that $\{\frac{7}{8}\}$ is a CLM set for $P(\mu)$ with respect to V. Now, we construct perturbations of $P(\mu)$ which have no local minimizers near $\frac{7}{8}$: Let $\{\mu_n\}_{n \ge 2}$ be the sequence in $\mathcal{P}(\mathbf{R})$ given by the uniform distributions on $[0, \frac{1}{4} + \frac{1}{n}]$. For $n \to \infty$ this sequence weakly converges to μ and we have

$$Q(x, \mu_n) = \begin{cases} 0 & 0 \le x \le \frac{3}{4} - \frac{1}{n} \\ -(4 - \frac{16}{n+4})x + 3 - \frac{16}{n+4} & \frac{3}{4} - \frac{1}{n} \le x \le 1 \\ -1 & 1 \le x \le \frac{7}{4} - \frac{1}{n}, \end{cases}$$

i. e. the objective of $P(\mu_n)$ is strictly increasing on $[\frac{3}{4}, 1]$ and no local minimizers occur near $\frac{7}{8}$.

As an application of Theorem 2.3 we now obtain the following stability result for two–stage stochastic integer programs. As in Section 2, the parameter space $\mathcal{P}(\mathbf{R}^s)$ is endowed with the topology of weak convergence.

Theorem 3.1 *Assume (2.1), (2.2) and let $\mu \in \Delta_{p,K}(\mathbf{R}^s)$ for some $p > 1$, $K > 0$. Let further μ be absolutely continuous with respect to the Lebesgue measure and suppose that $M \subset \mathbf{R}^m$ is a CLM set for $P(\mu)$ with respect to some bounded open set $V \subset \mathbf{R}^m$, i.e. $M = \psi_V(\mu)$.*
Then

(i) the function φ_V (from $\Delta_{p,K}(\mathbf{R}^s)$ to $\mathbf{R}$) is continuous at μ,

(ii) the multifunction ψ_V (from $\Delta_{p,K}(\mathbf{R}^s)$ to $\mathbf{R}^m$) is Berge upper semicontinuous at μ, i. e. for any open set G in $\mathbf{R}^m$ with $G \supset \psi_V(\mu)$ there exists a neighbourhood U of μ in $\Delta_{p,K}(\mathbf{R}^s)$ such that $\psi_V(\mu') \subset G$ whenever $\mu' \in U$,

(iii) there exists a neighbourhood U' of μ in $\Delta_{p,K}(\mathbf{R}^s)$ such that for all $\mu' \in U'$ we have that $\psi_V(\mu')$ is a CLM set for $P(\mu')$ with respect to V.

Proof: Thanks to Theorem 2.3, the proof of (i) and (ii) reduces to simplifications of arguments one usually has to give when establishing Berge's stability theory for more general parametric programs (cf. e. g. [1]).
For arbitrary $x \in \mathbf{R}^m$ and $\mu' \in \Delta_{p,K}(\mathbf{R}^s)$ denote

$$F(x, \mu') := f(x) + Q(x, \mu').$$

Consider $\mu_n \in \Delta_{p,K}(\mathbf{R}^s)$ such that $\mu_n \xrightarrow{w} \mu$.
Let $\{\alpha_n\}$ be a sequence of positive reals converging to zero. Then, there exist $x_n \in C \cap \operatorname{cl} V$ such that

$$\varphi_V(\mu_n) \leq F(x_n, \mu_n) \leq \varphi_V(\mu_n) + \alpha_n.$$

Since $C \cap \operatorname{cl} V$ is compact, we have an accumulation point x' of $\{x_n\}$ in that set. Passing to the limit in the above inequalities then yields, by continuity of F at (x', μ),

$$\limsup_{n \to \infty} \varphi_V(\mu_n) \leq F(x', \mu) \leq \liminf_{n \to \infty} \varphi_V(\mu_n).$$

Hence, $\lim_{n \to \infty} \varphi_V(\mu_n)$ exists and coincides with $F(x', \mu)$.
If there were $\alpha > 0$ and $x'' \in C \cap \operatorname{cl} V$ such that

$$F(x'', \mu) = F(x', \mu) - \alpha,$$

then continuity of F at (x'', μ) and $\lim_{n \to \infty} \varphi_V(\mu_n) = F(x', \mu)$ would imply that for n sufficiently large

$$F(x'', \mu_n) < F(x'', \mu) + \alpha/2 = F(x', \mu) - \alpha/2 < \varphi_V(\mu_n).$$

This contradiction verifies (i).

By compactness of $C \cap \operatorname{cl} V$ Berge upper semicontinuity of ψ_V at μ is equivalent to closedness of ψ_V at μ (cf. [1]), i. e. we have to establish that $\mu_n \overset{w}{\to} \mu$, $x_n \in \psi_V(\mu_n)$ and $x_n \to x'$ imply $x' \in \psi_V(\mu)$. This follows immediately from

$$F(x', \mu) = \lim_{n \to \infty} F(x_n, \mu_n) = \lim_{n \to \infty} \varphi_V(\mu_n) = \varphi_V(\mu).$$

To establish (iii) we show that $Q(., \mu')$ is lower semicontinuous for any $\mu' \in \Delta_{p,K}(\mathbf{R}^s)$. Then, compactness of $C \cap \operatorname{cl} V$ yields $\psi_V(\mu') \neq \emptyset$ and, by Berge upper semicontinuity of ψ_V, there exists a neighbourhood U' of μ in $\Delta_{p,K}(\mathbf{R}^s)$ such that $\psi_V(\mu') \subset V$ whenever $\mu' \in U'$, i. e. $\psi_V(\mu')$ is a CLM set for $P(\mu')$ with respect to V.

Let $x \in \mathbf{R}^m$ and $\{x_n\}$ be a sequence in $\mathbf{R}^m$ such that $x_n \to x$ for $n \to \infty$. With some $r > 0$, we have $\|x - x_n\| \leq r$ for all n. Furthermore, in view of Proposition 2.1,

$$
\begin{aligned}
\Phi(z - Ax_n) &\geq \Phi(z - Ax) - |\Phi(z - Ax_n) - \Phi(z - Ax)| \\
&\geq \Phi(z - Ax) - \alpha\|Ax_n - Ax\| - \beta \\
&\geq \Phi(z - Ax) - \alpha\|A\|r - \beta.
\end{aligned}
$$

The integral

$$\int_{\mathbf{R}^s} \Phi(z - Ax)\mu'(dz)$$

is finite since $\mu' \in \Delta_{p,K}(\mathbf{R}^s)$ (cf. the justification of assumption (2.3) or [12], Lemma 3.1).

Hence, the function

$$\tilde{h}(z) := \Phi(z - Ax) - \alpha\|A\|r - \beta$$

is integrable with respect to μ' and minorizes $\Phi(z - Ax_n)$ for all n.

Now, lower semicontinuity of Φ (cf. [4], p. 133) and Fatou's Lemma imply

$$Q(x, \mu') = \int_{\mathbf{R}^s} \Phi(z - Ax)\mu'(dz)$$

$$\leq \int_{\mathbf{R}^s} \liminf_{n\to\infty} \Phi(z - Ax_n)\mu'(dz)$$

$$\leq \liminf_{n\to\infty} \int_{\mathbf{R}^s} \Phi(z - Ax_n)\mu'(dz)$$

$$= \liminf_{n\to\infty} Q(x_n, \mu'),$$

which verifies the lower semicontinuity of $Q(., \mu')$.

q.e.d.

Note that in the above theorem we do not assume continuity of the objectives in the perturbed programs $P(\mu')$. In two–stage stochastic integer programming this lack of continuity typically arises when approximating the underlying probability measure by discrete ones (cf. Remark 2.4).
If both the original measure and the perturbed ones are absolutely continuous with respect to the Lebesgue measure, then the above theorem works if densities of the measures converge pointwise except for a set having Lebesgue measure zero, since the latter is a sufficient condition for weak convergence ([3] Scheffe's Theorem p. 224, Portmanteau Theorem p. 11).

Acknowledgements: This paper was written while visiting the University of Bergen, Norway, supported by a grant of Ruhrgas via NAVF. The author wishes to thank Sjur Flåm (University of Bergen) and Werner Römisch (Humboldt-University Berlin) for beneficial discussions. Further thanks are due to an anonymous referee for valuable comments.

References

[1] Bank, B.; Guddat, J.; Klatte, D.; Kummer, B.; Tammer, K.: Nonlinear Parametric Optimization, Akademie-Verlag, Berlin, 1982.

[2] Bank, B.; Mandel, R.: Parametric Integer Optimization, Akademie-Verlag, Berlin, 1988.

[3] Billingsley, P.: Convergence of Probability Measures, Wiley, New York, 1968.

[4] Blair, C. E.; Jeroslow, R. G.: The value function of a mixed integer program: I, Discrete Mathematics 19(1977), 121-138.

[5] Kall, P.: Stochastic Linear Programming, Springer-Verlag, Berlin, 1976.

[6] Kall, P.: Stochastic programming, European Journal of Operational Research 10(1982), 125-130.

[7] Kall, P.: On approximations and stability in stochastic programming, in: J. Guddat, H. Th. Jongen, B. Kummer, F. Nožička, eds., Parametric Optimization and Related Topics, Akademie-Verlag, Berlin, 1987, 387-407.

[8] Klatte, D.: A note on quantitative stability results in nonlinear optimization, in: K. Lommatzsch, ed., Proceedings of the 19. Jahrestagung Mathematische Optimierung, Seminarbericht Nr. 90, Humboldt-Universität Berlin, Sektion Mathematik, 1987, 77-86.

[9] Rinnooy Kan, A.; Stougie, L.: Stochastic integer programming, in: Y. Ermoliev, R. J. B. Wets, eds., Numerical Techniques for Stochastic Optimization, Springer-Verlag, Berlin, 1988, 201-213.

[10] Robinson, S. M.: Local epi-continuity and local optimization, Mathematical Programming 37(1987), 208-222.

[11] Robinson, S. M.; Wets, R. J. B.: Stability in two–stage stochastic programming, SIAM Journal on Control and Optimization 25(1987), 1409-1416.

[12] Schultz, R.: Continuity properties of expectation functionals in stochastic integer programming, manuscript, Humboldt-Universität Berlin, Sektion Mathematik, 1990.

[13] Stougie, L.: Design and analysis of algorithms for stochastic integer programming, CWI Tract 37, Centrum voor Wiskunde en Informatica, Amsterdam, 1987.

[14] Wets, R. J. B.: Stochastic programming: solution techniques and approximation schemes, in: A. Bachem, M. Grötschel, B. Korte, eds., Mathematical Programming: The-State-of-the-Art 1982, Springer-Verlag, Berlin, 1983, 566-603.

THREE APPROACHES FOR SOLVING THE STOCHASTIC
MULTIOBJECTIVE PROGRAMMING PROBLEM

Norio Baba
Osaka Educational University, Ikeda City,
563, Japan.

Akira Morimoto
Kyoto University, Kyoto City, 606, Japan.

Abstract.   In this paper, we consider the multiobjective optimization
problem in which each objective function is disturbed by noise.   Three
approaches using learning automata, random optimization method, and stochastic
approximation method are proposed to solve this problem.   It is shown that
these three approaches are able to find appropriate solutions of this problem.
Several computer simulation results also confirm our theoretical study.

I. INTRODUCTION

In order to find an appropriate solution in a situation with several conflicting
objectives (goals), multiobjective optimization problems have been extensively studied
by many researchers and various optimization techniques have been developed ([1] $\sim$
[4]).   Almost all of the researches in this area have so far been done under the
assumption that value of each objective function can be measured without an error.
However, we often encounter real situations in which this assumption does not hold.
(Real systems are often disturbed by noise.)
In this paper, we consider the stochastic multiobjective optimization problem in
which the value of each objective function can be measured only through additive noise.
The following three approaches for solving this problem are proposed in this paper:

1)   An approach using learning automata
2)   An approach using random optimization method
3)   An approach using stochastic approximation method

Convergence properties of the above three approaches are considered.
The organization of this paper is as follows.   The first part is devoted to the
problem statement.   Here, the problem and notations are defined.   In the second part,
a learning automaton [5], [6] is proposed to solve this problem.   It is shown that an
appropriately chosen learning scheme of variable-structure stochastic automaton ensures
convergence to reasonable solutions among the finite candidates of the solutions.

In the third part, an approach using random optimization method of Matyas [7] (or its modified version) is proposed. It is shown that this approach ensures convergence with probability 1 to an arbitrary neighborhood of satisficing Pareto-optimal solutions under several assumptions. In the fourth part, a stochastic approximation method is proposed. Convergence of the proposed method is also discussed. The final part is devoted to the concluding remarks. Here, merits and demerits of the three approaches are discussed.

II.   STATEMENT OF THE PROBLEM

In this paper, we consider the following stochastic multiobjective optimization problem (P):

(P)
$$\text{Maximize}\ \ f(x), \qquad f(x) = \begin{pmatrix} f_1(x) \\ \vdots \\ f_r(x) \end{pmatrix} \tag{1}$$

, where $f:\ R^n \longrightarrow R^r$ is a vector-valued objective function.

Informations concerning $f(x)$ can only be obtained from the following observations disturbed by noise:

$$y_i(x)\ =\ f_i(x) + \eta_i(x) \qquad (\ i = 1,\ldots,r\ ) \tag{2}$$

, where $\eta_i(x)$ is an additive noise.

In the ordinary (deterministic) multiobjective optimization problem in which $\eta_i(x) = 0$ in (2), the definition of Pareto-optimal solution is given as follows:

Definition 1:   Let $\hat{x} \in X$.   If we cannot find any x that satisfies the inequality $f(x) \geq f(\hat{x})$, $x \in X$, then $\hat{x}$ is said to be Pareto-optimal solution. (Here, $f(x) \geq f(\hat{x})$ means that $f_i(x) \geq f_i(\hat{x})$ for all i, $1 \leq i \leq r$ and $f(x) \neq f(\hat{x})$.)

Suppose that we are given an aspiration level $\overline{f} = (\ \overline{f}_1,\ldots,\overline{f}_r\ )$ so that the solution $\tilde{x}$ of (1) should satisfy

$$f_i(\tilde{x})\ \geq\ \overline{f}_i, \qquad i = 1,\ldots,r \tag{3}$$

Then, our objective is to find a Pareto-optimal solution which satisfies (3). However, in the problem (P), we can measure value of each objective function $f_i(x)$ only through additive noise.

In the following sections, we shall propose three approaches in order to find an appropriate solution of (P).

Remark 2.1:   In this paper, there are no recommendations for choosing aspiration level $\overline{f}$.   We assume that one can choose an appropriate aspiration level for each stochastic multiobjective programming problem.

Remark 2.2:   We shall not give any special definition in dealing with the stochastic multiobjective optimization problem (P).   Instead, we shall use same definitions which are given in the deterministic multiobjective optimization problem.

III.   AN APPROACH USING LEARNING AUTOMATA

In this section, we propose an approach using learning automata.   First of all, let us briefly introduce basic concepts of learning automata.

The learning mechanism of stochastic automaton A operating in the general nonstationary multiteacher environment is described in Figure 1.   The stochastic automaton A is defined by the 6-tuple ($S,W,O,g,P(t),T$).   S is a set of random variables s_j^i with values in [0,1]: $S = \{ s_1^i,\ldots,s_r^i \}$ where s_j^i (j=1,\ldots,r) is the response from the jth teacher R_j (j=1,\ldots,r).

W denotes the set of $\overline{r}$ internal states $\{w_1,\ldots,w_{\overline{r}}\}$.   O denotes the set of $\overline{r}$ outputs $\{ o_1,\ldots,o_{\overline{r}} \}$.   g denotes the output function o(t) = g(w(t)), that is a

Figure 1:   Nonstationary Multiteacher Environment NMT

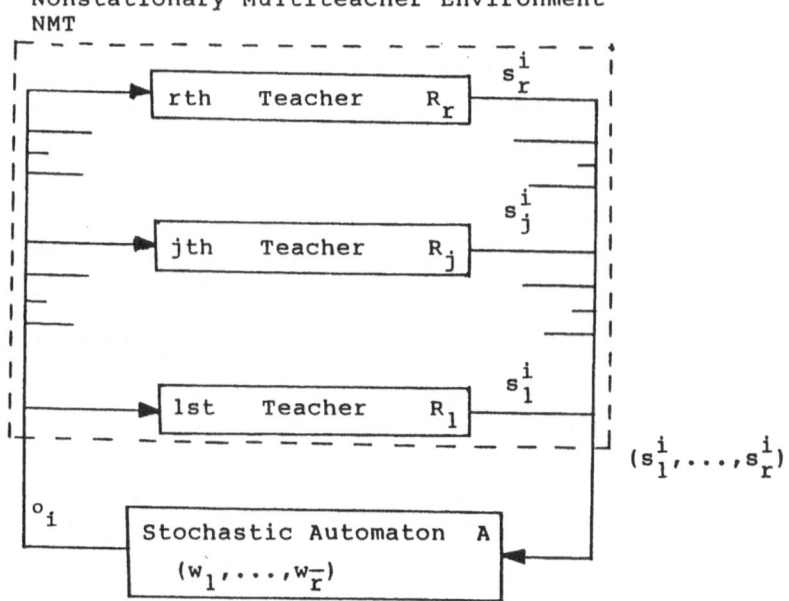

one-to-one deterministic mapping. $P(t)$ denotes the probability vector ($P(t) = (p_1(t),\ldots,p_{\overline{r}}(t))'$) at time t which governs the choice of the state. T denotes the reinforcement scheme which generates $P(t+1)$ from $P(t)$. The state w_k ($k=1,\ldots,\overline{r}$) is chosen at time t with probability $p_k(t)$, where

$$p_1(0) = \ldots = p_{\overline{r}}(0) = 1/\overline{r}, \qquad \sum_{i=1}^{r} p_i(t) = 1.$$

Assume now that the state w_i is chosen at time t. Then, the stochastic automaton A performs action o_i on the nonstationary multiteacher environment (NMT). In response to o_i, the jth teacher R_j emits output s_j^i. Here, s_j^i is a function of t and ω. ($\omega \in \Omega$; Ω is the basic ω-space of the probability measure space (Ω,B,μ), and B is the smallest Borel field including $\bigcup_{t=0}^{\infty} F_t$, where $F_t = \sigma(\,P(u),\ s_j^i(u)\ :\ (u=0,\ldots,t)\,)$. Consequently, from now on, we shall use the notation $s_j^i(t,\omega)$ to represent the input to the stochastic automaton A. Depending upon the action o_i and the r responses from r teachers, the stochastic automaton A changes the probability vector $P(t)$ by the reinforcement scheme T.

In order to tackle the stochastic multiobjective optimization problem (P) by using the learning performance of the stochastic automaton, we have to reformulate the original problem as follows.

Suppose that there is a finite number of candidates of the solution of problem (P). Let $\{ x_1,\ldots,x_{\overline{r}} \}$ be the set of these candidates. Then, the problem here is to find the most appropriate candidate x_α.

Let us try to identify the ith action o_i of the stochastic automaton A with the ith parameter value x_i ($i = 1,\ldots,\overline{r}$). That is to say, choosing the ith parameter x_i at time t corresponds to A producing output o_i at time t.

Then, the original problem can be reduced to the problem in which we have to find an appropriate parameter x_α among the $\overline{r}$ parameters $x_1,\ldots,x_{\overline{r}}$ by using the learning performance of the stochastic automaton A.

The MGAE scheme was recently proposed by one of the authors [9]. It was shown that this scheme ensures ε-optimality under the condition (C):

MGAE Scheme   Suppose that $y(t) = y_i$ and the responses from NMT are $s_1^i,\ldots,s_r^i$. Then,

$$p_i(t+1) = p_i(t) + [\frac{s_1^i+\ldots+s_r^i}{r}]\{ \sum_{j\neq i}^{\overline{r}} \phi_j(P(t)) \} - [1 - \frac{s_1^i+\ldots+s_r^i}{r}]\{ \sum_{j\neq i}^{\overline{r}} \psi_j(P(t)) \}$$

$$p_j(t+1) = p_j(t) - [\frac{s_1^i+\ldots+s_r^i}{r}]\{ \phi_j(p(t)) \} + [1 - \frac{s_1^i+\ldots+s_r^i}{r}]\{ \psi_j(P(t)) \} \qquad (j\neq i)$$

where ϕ_i, ψ_i $(i=1,\dots,\bar{r})$ satisfy the following relations:

$$\frac{\phi_1(P(t))}{p_1(t)} = \frac{\phi_2(P(t))}{p_2(t)} = \ \dots\ = \frac{\phi_{\bar{r}}(P(t))}{p_{\bar{r}}(t)} = \lambda(P(t))$$

$$\frac{\psi_1(P(t))}{p_1(t)} = \frac{\psi_2(P(t))}{p_2(t)} = \ \dots\ = \frac{\psi_{\bar{r}}(P(t))}{p_{\bar{r}}(t)} = \mu(P(t))$$

($\lambda(P(t))$ and $\mu(P(t))$ are nonlinear functions of the components of the vector $P(t)$.)

$$p_j(t) + \psi_j(P(t)) > 0. \qquad\qquad p_i(t) + \sum_{\substack{j=1\\ j\neq i}}^{\bar{r}} \phi_j(P(t)) > 0.$$

$$p_j(t) - \phi_j(P(t)) < 1 \qquad (j = 1,\dots,\bar{r} \ ; \ i = 1,\dots,\bar{r})$$

Remark 3.1:   "MGAE scheme" is an abbreviation of "Absolutely Expedient Reinforcement Scheme in the General Multiteacher Environment". [6]

Assume that the nonstationary multiteacher environment NMT has the following property:

(C) $$\int_0^1 s\,dF_{\alpha,t}(s) + \frac{\delta}{r} < \int_0^1 s\,dF_{j,t}(s) ,\qquad\qquad (4)$$

where $F_{i,t}(s)$ $(i=1,\dots,\bar{r})$ is the distribution function of $(s_1^i(t,\omega)+\dots+ s_r^i(t,\omega))/r$, for some state w_α, some $\delta > 0$, all time t, all j ($\neq \alpha$), and all ω ($\in \Omega$).

Then, it can be easily shown [9] that the following theorem holds:

THEOREM: Suppose that $\lambda(P(t)) = \theta\{\bar{\lambda}(P(t))\}$ ($\theta > 0$)   and $\mu(P(t)) = \theta\{\bar{\mu}(P(t))\}$, where $\bar{\lambda}(P(t))$ and $\bar{\mu}(P(t))$ are bounded functions which satisfy the following conditions:

$\bar{\lambda}(P(t)) \leq 0$, $\bar{\mu}(P(t)) \leq 0$, and $\bar{\lambda}(P(t)) + \bar{\mu}(P(t)) < 0$, for all $P(t)$ and t.

Then, the stochastic automaton A with the MGAE reinforcement scheme is ε-optimal under the nonstationary multiteacher environment satisfying the condition (C). That is to say, one can choose (for any $\varepsilon > 0$) a parameter θ included in the reinforcement scheme of the stochastic automaton A such that the following equality holds:

$$\lim_{t\to\infty} E\{p_\alpha(t)\} \geq 1 - \varepsilon$$

<u>Remark 3.2</u>: s_j^i ($i = 1,\dots,\bar{r}$; $j = 1,\dots,r$) denotes penalty strength.
The inequality (4) means that the output o_α receives the least average penalty
strength from the r teacher environment among the $\bar{r}$ outputs.   Therefore, the above
theorem indicates that the stochastic automaton with MGAE scheme ensures convergence
to the most resonable output.

In order to utilize the learning performance of the MGAE scheme for our stochastic
multiobjective optimization problem, the following penalty strength s_j^i (which has to
be an input into stochastic automaton A) is assigned:

$$s_j^i \;=\; \text{Max}\;(\;s_j^i(a),\; s_j^i(b)\;)\qquad (\;i = 1,\dots,\bar{r}\;;\; j = 1,\dots,r\;) \qquad (5)$$

where $s_j^i(a)$ and $s_j^i(b)$ are obtained by the calculation as shown in Figure 2 and
Figure 3, respectively.

Now, let us assume that the stochastic automaton A with the MGAE reinforcement
scheme has been applied to find an appropriate solution of the stochastic multiobjective optimization problem (P).   Then, from the analogy with the above theorem,
we can easily get:

<u>THEOREM</u>:   Assume that there exists at least one Pareto-optimal solution
satisfying the condition (3) among $x_1,\dots,x_{\bar{r}}$.   Let $p_{\delta_1}(t)$ be the sum of the
state probabilities corresponding to the Pareto-optimal solutions satisfying the
condition (3).   Further, let $p_{\delta_2}(t)$ be the sum of the state probabilities whose
corresponding solutions are not   completely inferior to one of the Pareto-optimal
solutions satisfying the condition (3).

Let $p_\delta(t) \;=\; p_{\delta_1}(t) + p_{\delta_2}(t)$

Then, it can be shown that $\lim\limits_{\theta\to 0}\lim\limits_{t\to\infty} E\{p_\delta(t)\} \;=\; 1$ \qquad (6)

<u>Remark 3.3</u>:   "Not completely inferior to one of the Pareto-optimal solutions"
means: Let $x_{i_1},\dots,x_{i_q}$ be the Pareto-optimal solutions.   Then, a solution x_γ
is said to be "not completely inferior to one of the Pareto-optimal solutions" if
there exists some components x_{i_s} ($1 \le s \le q$) and $f_k(x_\gamma) > f_k(x_{i_s})$ for some
k ($1 \le k \le r$).

<u>Remark 3.4</u>: θ denotes a parameter included in $\phi_i(P(t))$ or $\psi_i(P(t))$. (i=1,
$\dots,\bar{r}$)   Consequently, $E\{p_\delta(t)\}$ in (6) includes parameter θ.

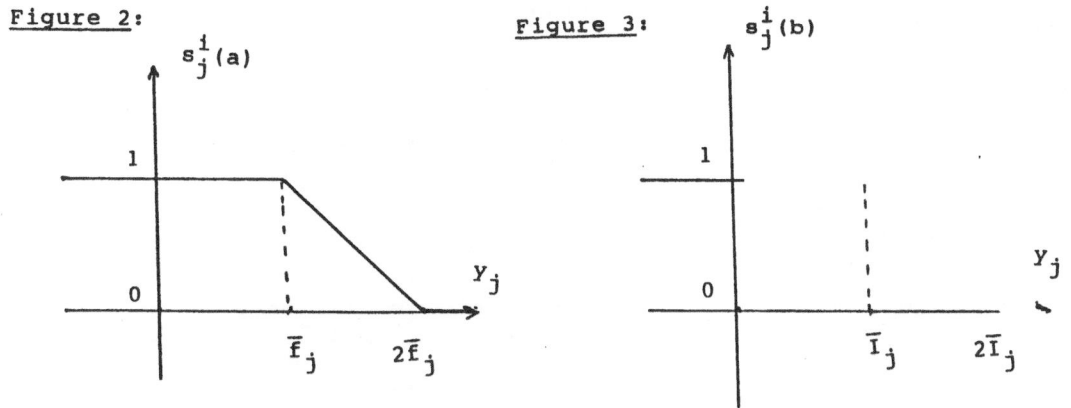

Figure 2:

$s_j^i(a)$

Figure 3:

$s_j^i(b)$

$\overline{I}_j$: Average value of y_j having been measured.

COMPUTER SIMULATION RESULT

In the following, we present a computer simulation result which illustrate learning performance of the MGAE reinforcement scheme under the noise corrupted multi-objective optimization problem.

The objective functions considered in this simulation are:

$$f_1(x,y) = -0.7x^2 - (y+1)^2 + 10$$
$$f_2(x,y) = -x^4/4 - x^3/6 + x^2/2 - (y-0.5)^2 + 7$$

Variance of noise $\eta_i(x)$ $(i=1,2)$:   0.2 $(\sigma_1(x))$ 0.1 $(\sigma_2(x))$

Aspiration level: $\overline{f}_1 = 5.0$ $\overline{f}_2 = 3.0$

Figure 4 to Figure 6 present the graphic displays appeared on the screen of the personal computer. Figure 4 illustrates the Pareto-optimal solutions satisfying the condition (3). Figure 5 illustrates the changes of the probabilities of the states of stochastic automaton. The states corresponding to the Pareto-optimal solutions satisfying the condition (3) are x_8, x_9, x_{12}, and x_{13}. Figure 6 illustrates the changes of the sum of the probabilities $p_8(t)$, $p_9(t)$, $p_{12}(t)$, and $p_{13}(t)$. The computer simulation result shown in Figure 5 and Figure 6 confirms our theoretical study.

Figure 4:   Figure 5:

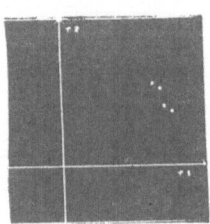 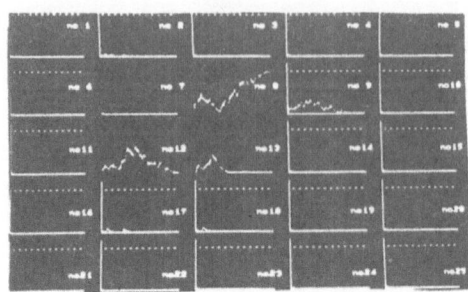

Figure 6:

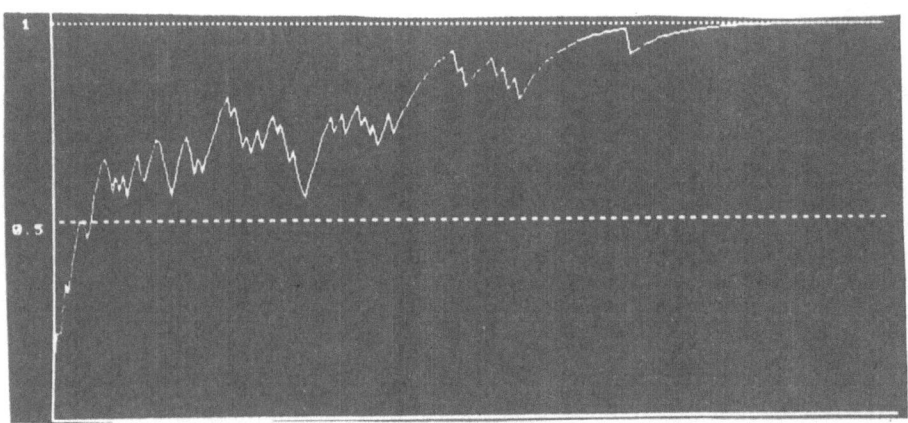

IV. AN APPROACH USING RANDOM OPTIMIZATION METHOD

In order to solve the stochastic multiobjective optimization problem (P), we shall propose an approach using random optimization method [8].

Let $\bar{f} = (\bar{f}_1, \ldots, \bar{f}_r)$ be the aspiration level of each objective function.

<u>Step 1.</u>   Choose an initial point $x^{(1)}$ in X. Let $\bar{a}_1^1 = y_1(x^{(1)}), \ldots,$
$\bar{a}_r^1 = y_r(x^{(1)})$. Further, let $\bar{a}^1 = (\bar{a}_1^1, \ldots, \bar{a}_r^1)$ and $k = 1$.

Step 2.   Generate Gaussian random vector $\xi^{(k)}$ from the sample space $(\, R^n, B, \mu_k\,)$.   Calculate $\tilde{x}_{k+1}$ from some generation function $D_i(x^{(k)}, \xi^{(k)})$.

If $\tilde{x}_{k+1}$ satisfies one of the following three conditions a),b), and c), then let $x^{(k+1)} = \tilde{x}_{k+1}$ and $\bar{a}_i^{k+1} = y_i(\tilde{x}_{k+1})$ ($i = 1,\ldots,r$).   Otherwise, let $x^{(k+1)} = x^{(k)}$ and $\bar{a}^{k+1} = \bar{a}^k$.

a) $y(\tilde{x}_{k+1}) \geq \bar{f}$, $\bar{a}^k \geq \bar{f}$, $y(\tilde{x}_{k+1}) \geq \bar{a}^k$

b) $y(\tilde{x}_{k+1}) \geq \bar{f}$, $\bar{a}^k \ngeq \bar{f}$

c) $y(\tilde{x}_{k+1}) \ngeq \bar{f}$, $y(\tilde{x}_{k+1}) \geq \bar{a}^k$

Step 3.   Let $k = k + 1$ and go back to Step 2.

Remark 4.1:   The above algorithm can be easily obtained from the random optimization methods. [7],[8]

Remark 4.2:   Examples of $D_i(x^{(k)}, \xi^{(k)})$.

1) $D_i(x^{(k)}, \xi^{(k)}) = x^{(k)} + \xi^{(k)}$

2) $D_i(x^{(k)}, \xi^{(k)}) = \begin{cases} x^{(k)} + b^{(k)} + \xi^{(k)} & (\,b^{(1)} = 0\,) \qquad (7) \\ x^{(k)} - b^{(k)} - \xi^{(k)} & \qquad\qquad\qquad\quad (8) \end{cases}$

If (7) satisfies one of the conditions a) to c), then let $b^{(k+1)} = 0.4\xi^{(k)} + 0.2b^{(k)}$.   If (8) satisfies one of the conditions a) to c), then let $b^{(k+1)} = b^{(k)} - 0.4\xi^{(k)}$.   If (8) does not satisfy any one of the three conditions, then let $b^{(k+1)} = 0.5b^{(k)}$.

In the above random optimization method, $x^{(k)}$ is updated by $\tilde{x}_{k+1}$ if $y(\tilde{x}_{k+1})$ assumed a better value than $\bar{a}^k$ which has been the best value among those having been obtained so far.

The following theorem shows that the above algorithm ensures convergence to the Pareto-optimal solutions satisfying the condition (3) under several assumptions:

<u>Theorem</u>: Let $f_i(x)$ $(i=1,\ldots,r)$ be uniformly continuous. Let $F_{\bar{f}}^P$ be the region of f-space corresponding to the Pareto-optimal solutions satisfying the condition (3).

Further, let $F_{\bar{f}}^{P,\varepsilon} = \{\; f \mid |f_i - \hat{f}_i| \le \varepsilon \text{ for all } i \; (1 \le i \le r) \text{ and some } \hat{f} \text{ in } F_{\bar{f}}^P \;\}$

$X_{\bar{f}}^{P,\varepsilon} = \{\; x \mid x \in X, \; f(x) \in F_{\bar{f}}^{P,\varepsilon} \;\}$

Assume that the following conditions i),ii),iii) and iv) hold:

i) $F_{\bar{f}}^P \ne \phi$

ii) Let $q_i(\cdot)$ be the probability density function of $\eta_i(x)$. Then,

$$m := \inf_{x \in X,\; 1 \le i \le r} q_i(x) > 0 \tag{9}$$

iii) $$M_i := \sup_{x \in X} \eta_i(x) < \infty \tag{10}$$

iv) The Gaussian random vector $\xi^{(k)}$ has the probability density function $p_k(x)$. There exists some positive number b which satisfies

$$b := \inf_{x \in X} p_k(x) \tag{11}$$

Then, the proposed algorithm ensures $\lim\limits_{k \to \infty} P\{\; x^{(k)} \in X_{\bar{f}}^{P,\varepsilon} \;\} = 1$ for an arbitrary chosen positive number ε.

<u>Remark 4.3</u>: The conditions of the above theorem are too strict. Therefore, future research effort is needed to mitigate conditions of the above theorem. In the following, we show an example that satisfies all of the above conditions:

The region $X = \{\; x \mid g(x) \le 0 \;\}$ is bounded in the sense that there exists some positive number J that satisfies $\| x \| < J$ for all $x \in X$. Further, each component of $\xi^{(k)}$ is generated from Gaussian distributions with positive variances.

The proof of the above theorem can be easily obtained:

<u>Proof</u>: Let us consider the probability of the event:

$$\overline{a}_i^k \ge f_i(\hat{x}) + M_i - \varepsilon \quad (1 \le i \le r \; ; \; \hat{f}_i = f_i(\hat{x})\;) \quad \text{for some } \hat{f} \text{ in } F_{\bar{f}}^P \tag{12}$$

Since $f_i(x)$ is a uniformly continuous function, there exists some $\delta_i > 0$ for any $\epsilon/2 > 0$ such that the following relation holds. ($i = 1,\dots,r$)

$$|| \, x - \hat{x} \, || \; < \; \delta_i \quad \longrightarrow \quad |\, f_i(x) - f_i(\hat{x}) \,| \; < \; \epsilon/2 \tag{13}$$

Let $\delta = \min(\delta_1,\dots,\delta_r)$. Then, from (13), we can get:

$$|| \, x - \hat{x} \, || \; < \; \delta \quad \longrightarrow \quad |\, f_i(x) - f_i(\hat{x}) \,| \; < \; \epsilon/2 \qquad \text{for all } i, \; 1 \leq i \leq r.$$

Further, let s be the Euclidean measure of

$$U_\delta(\hat{x}) \; = \; \{ \, x \mid \; || \, x - \hat{x} \, || \, < \, \delta, \; x \in X \, \} \tag{14}$$

Then, from the algorithm and (11), we can get:

$$P(\, \tilde{x}_k \in X_{\bar{f}}^{P,\epsilon/2} \,) \; \geq \; sb \tag{15}$$

Moreover, the probability that $y(\tilde{x}_k)$ satisfies (12) is larger than $(\frac{\epsilon}{2}m)^r$. (16)

Therefore, assuming that $\overrightarrow{a}^k$ does not satisfy (12), the probability that $\overrightarrow{a}^{k+1}$ satisfies (12) in the next step is larger than $sb(\frac{\epsilon}{2}m)^r$ ($\because$ (15), (16))

Let $\beta = sb(\frac{\epsilon}{2}m)^r$.

Assume that $\overrightarrow{a}^k$ does not satisfy (12).

Once $\tilde{x}_{k+1} \in X_{\bar{f}}^{P,\epsilon/2}$ and (12) is satisfied at $k = k+1$, $x^{(t)}$ ($t=t+2, t+3,\dots$) cannot escape from $X_{\bar{f}}^{P,\epsilon}$.

Hence,
$$P(\, x^{(t)} \in X_{\bar{f}}^{P,\epsilon} \,) \; \geq \; P(\, \tilde{x}_{k+1} \in X_{\bar{f}}^{P,\epsilon/2} \text{ and } \overline{a}_i^{k+1} \geq f_i + M_i - \epsilon, \; 1 \leq i \leq r \,)$$
$$\geq \; 1 - (1 - \beta)^k \qquad (\, t = k+2, k+3,\dots \,)$$

$$\therefore \quad \lim_{t \to \infty} P(\, x^{(t)} \in X_{\bar{f}}^{P,\epsilon} \,) \; \geq \; \lim_{k \to \infty} \{ 1 - (1 - \beta)^k \} \; = \; 1$$

COMPUTER SIMULATION RESULT

In the following, we shall give a computer simulation result which was recently carried out by using the proposed random optimization method.

<u>Objective functions:</u> $f_1(x,y) = -100(x^2 + y^2) - (1 - x)^2,$

$f_2(x,y) = -(y - 5.1x^2/(4/\pi)^2 + 5x/\pi - 6)^2,$ $f_3(x,y) = -4x^2 + 2.1x^4 - x^6/3 - xy$
$+ 4y^2 - 4y^4,$ $f_4(x,y) = -0.03x^4 + 1.2xy - y^2.$

<u>Aspiration levels:</u> $\overline{f}_1 = -100,$ $\overline{f}_2 = -30,$ $\overline{f}_3 = -50,$ $\overline{f}_4 = -10.$

<u>Initial point:</u> $x = 10,$ $y = 10.$

<u>Total number of steps:</u>   500

k	x	y	f_1 (y_1)	f_2 (y_2)	f_3 (y_3)	f_4 (y_4)
0	10.00	10.00	-2.01E+04	-8.68E+04	-3.52E+05	-2.80E+02
1	6.38	10.55	-1.52E+04	-1.28E+04	-6.83E+04	-8.02E+01
			(-1.52E+04)	(-1.28E+04)	(-6.83E+04)	(-7.92E+01)
3	5.44	9.92	-1.28E+04	-6.46E+03	-4.52E+04	-5.99E+01
			(-1.28E+04)	(-6.47E+03)	(-4.52E+04)	(-6.16E+01)
6	4.78	9.59	-1.15E+04	-3.68E+03	-3.65E+04	-5.26E+01
			(-1.15E+04)	(-3.68E+03)	(-3.65E+04)	(-5.36E+01)
7	4.49	9.01	-1.02E+04	-2.85E+03	-2.81E+04	-4.49E+01
			(-1.01E+04)	(-2.85E+03)	(-2.80E+04)	(-4.50E+01)
11	4.15	8.81	-9.51E+03	-2.01E+03	-2.50E+04	-4.27E+01
			(-9.50E+03)	(-2.02E+03)	(-2.50E+04)	(-4.38E+01)
19	3.88	7.87	-7.70E+03	-1.54E+03	-1.58E+04	-3.21E+01
			(-7.71E+03)	(-1.55E+03)	(-1.58E+04)	(-3.11E+01)
27	3.52	6.60	-5.60E+03	-1.08E+03	-7.78E+03	-2.02E+01
			(-5.60E+03)	(-1.08E+03)	(-7.79E+03)	(-1.83E+01)
30	1.77	4.87	-2.68E+03	-6.67E+01	-2.16E+03	-1.37E+01
			(-2.69E+03)	(-6.37E+01)	(-2.17E+03)	(-1.24E+01)
47	1.07	3.39	-1.26E+03	-2.04E+01	-4.86E+02	-7.16E+00
			(-1.26E+03)	(-2.31E+01)	(-4.92E+02)	(-1.03E+01)
49	0.412	2.56	-6.73E+02	-1.10E+01	-1.47E+02	-5.29E+00
			(-6.80E+02)	(-6.83E+00)	(-1.43E+02)	(-2.87E+00)
78	0.0106	0.823	-6.87E+01	-2.66E+01	8.65E-01	-6.67E-01
			(-6.30E+01)	(-2.73E+01)	(6.37E+00)	(-3.67E+00)
473	0.0616	0.647	-4.31E+01	-2.77E+01	9.18E-01	-3.71E-01
			(-4.66E+01)	(-2.37E+01)	(8.77E+00)	(2.56E+00)

<u>Remark 4.4:</u>   As $\eta_i(x)$ $(i=1,2,3,4)$, we have used random variables with uniform distribution with variances $\sigma_1 = 5.0,$ $\sigma_2 = 3.0,$ $\sigma_3 = 5.0,$ $\sigma_4 = 2.0.$

<u>Remark 4.5:</u>   The values in the parentheses of the Table indicate the amount of each objective function disturbed by noise $\eta_i(x)$.

V. AN APPROACH USING STOCHASTIC APPROXIMATION METHOD

Stochastic approximation methods have been frequently used to solve various stochastic optimization problems. [11]$\sim$[17] In the following, we propose a stochastic approximation method for solving our stochastic multiobjective optimization problem (P). Assume that we are given a utopia point f^{*} and an aspiration point $\bar{f}$. (Here, f^{*}_{i} is assumed to be larger than $\max_{x \in X} f_{i}(x)$ (i=1,...,r))

Let $\pi_{X}(\cdot)$ denote a projection on the set X. Then, our algorithm can be described as follows:

$$x^{k+1} = \pi_{X}(x^{k} - \rho_{k}T^{k}) \tag{17}$$

$$T^{k} = \frac{1}{\alpha_{k}} \sum_{j=1}^{n} [F(\tilde{x}^{k}_{1},...,x^{k}_{j} + \frac{\alpha_{k}}{2},...,\tilde{x}^{k}_{n}) - F(\tilde{x}^{k}_{1},...,x^{k}_{j} - \frac{\alpha_{k}}{2},...,\tilde{x}^{k}_{n})] e_{j} \tag{18}$$

$$\tilde{x}^{k}_{j} = x^{k}_{j} + h(k,j), \qquad (j=1,...,n).$$

where

$$F(x) = \text{Max}_{i} \frac{f^{*}_{i} - y_{i}(x)}{f^{*}_{i} - \bar{f}_{i}} , \quad e_{j} \text{ is the unit vector on the jth axis, and}$$

$h(k,j)$ is a random variable with uniform distribution on $[-\alpha_{k}/2, \alpha_{k}/2]$.

Our objective is to find a Pareto-optimal solution which satisfies the condition (3). Wierzbicki et al [3],[4] proposed the approach for finding a Pareto-optimal solution satisfying the condition (3) in the usual (deterministic) multiobjective optimization problems. ($\eta_{i}(x) = 0$ in (P).) The above algorithm is originated from one of those approaches [4].

Concerning the convergence of the above algorithm, we can easily derive the following theorem:

THEOREM: Assume the followings:

1) $f_{i}(x)$ (i=1,...,r) is concave and continuous on an open set $\tilde{X} \subset X$

2) $f_{i}(x)$ is a local Lipschitz function in $\tilde{X}$.

3) X is a convex and compact set.

106

4) $\rho_k \downarrow 0$, $\alpha_k \downarrow 0$, $\sum\limits_{k=0}^{\infty} \rho_k = \infty$, $\sum\limits_{k=0}^{\infty} \rho_k^2 < \infty$, $\dfrac{\rho_k}{\alpha_k} \downarrow 0$, $\dfrac{\alpha_k - \alpha_{k+1}}{\alpha_k \rho_k} \to 0$,

$\sum\limits_{k=0}^{\infty} (\dfrac{\rho_k}{\alpha_k})^2 < \infty$

5) In our problem (P), the noise $\eta_i(x)$ is restricted to any bounded noise which satisfies $E\{\eta_i\} = 0$. ($i = 1,\ldots,n$)

Then, x^k converges with probability 1 to the global minimum of

$$f(x) = \underset{i}{Max} \frac{f_i^* - f_i(x)}{f_i^* - \overline{f}_i} . \tag{19}$$

Since the above theorem can be easily proved by using the same procedure as utilized in [15], we only describe the outline of the proof:

(1) Since $f_i(x)$ is concave and continuous, $f(x)$ is convex and continuous. But $f(x)$ does not have continuous derivatives.

(2) Then, we approximate $f(x)$ by the family of C^1 functions $F(x,k)$.

$$F(x,k) = \frac{1}{\alpha_k^n} \int_{-\epsilon}^{\epsilon} \cdots \int_{-\epsilon}^{\epsilon} f(x+y)\, dy_1 \ldots dy_n , \text{ where } \epsilon = \alpha_k/2 \ (k=1,2,\ldots).$$

(3) It can be easily shown that $F(x,k)$ uniformly converges to $f(x)$ in X as $k \to \infty$ and that T^k is the stochastic quasigradient of $F(x,k)$.

(4) By using the Theorem 4 in Section 6 of [15], we can easily prove the theorem.

Remark 5.1: Figure 7 illustrates the point of $\underset{i}{Max} \dfrac{f_i^* - f_i(x)}{f_i^* - \overline{f}_i}$.

Figure 7:

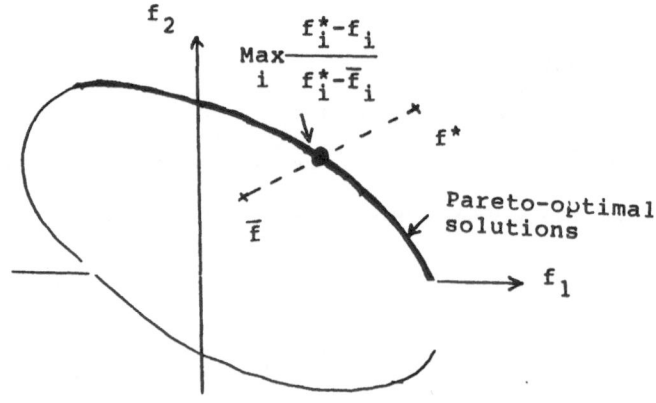

Remark 5.2:   In order to ensure convergence to the global minimum, f_i (i=1,..., r) has been assumed to be concave.   Without this assumption, the above algorithm can be trapped in a local minimum of (19).

Remark 5.3:   Some of the readers might have the question "Why is it not shown how the proposed algorithms can discover the emptiness of the Pareto-optimal solutions satisfying the condition (3) ?"   We can answer this question easily.

The first algorithm can find an appropriate solution from finite candidates of the solutions even though Pareto-optimal solution set satisfying the condition (3) is empty.   The convergence theorem concerning the second algorithm assumes the non-emptiness of the Pareto-optimal solution satisfying the condition (3).   Therefore, we cannot ensure any convergence property when we apply the second algorithm to the problem in which the nonemptiness of the Pareto-optimal solution satisfying the condi-tion (3) is not assumed.   The third algorithm does not need such an assumption. Even if $F_{\bar{f}}^P = \phi$, one can obtain one of the Pareto-optimal solutions.

COMPUTER SIMULATION RESULT

The proposed stochastic approximation method has been tested by computer simula-tions.   In the following, we give an example of those simulations:

Objective functions: $f_1(x,y) = -x^2 - (y - 4)^2 + 70,$

$f_2(x,y) = -x^2 - 0.8(y + 4)^2 + 65$

Aspiration level: $\overline{f}_1 = 25, \quad \overline{f}_2 = 35$

Utopia point: $f_1^* = 85, \quad f_2^* = 85$

Figure 8:

Initial point: x=9, y=9

Variance of noise: $\sigma_1 = 10,$
$\sigma_2 = 11.$

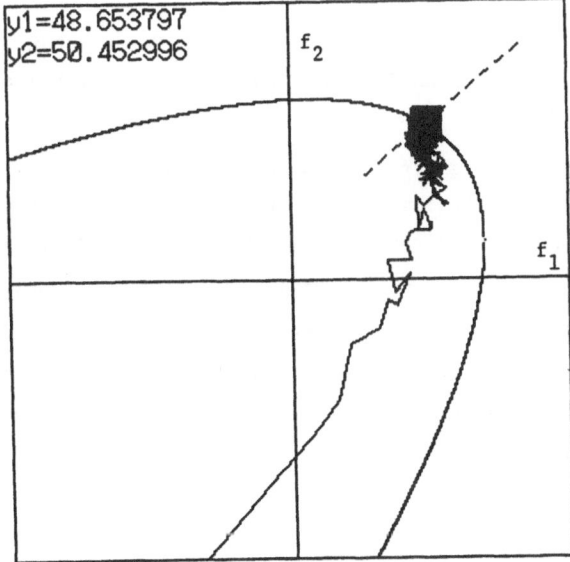

V. CONCLUSION

Three approaches for solving the stochastic multiobjective optimization problem have been proposed.   However, each approach has his own problem to be solved.   In the approach using learning automata, one is able to compare only finite candidates of solutions.   In the random optimization method, we need several conditions in order to ensure convergence to an appropriate solution.   Moreover, some of those conditions are too strict to be satisfied.   Although the stochastic approximation method seems to be most promising, we also need rather strict condition that $\eta_i(x)$ should be bounded noise.   (Therefore, any convergence property cannot be theoretically ensured when $\eta_i(x)$ happens to be Gaussian noise.)   The future research effort is needed to solve these problems.

In this paper, we have assumed that aspiration level $\overline{f}_i$ (i=1,...,r) is a priori given for each objective function.   However, many researchers suppose that finding aspiration level $\overline{f}_i$ (i=1,...,r) is the essence of the multiobjective optimization problem.   Therefore, future research effort should also be directed in order to derive a suitable method which introduces a scheme for determining $\overline{f}_i$ (i=1,...,r) into our proposed algorithm.

ACKNOWLEDGEMENT

The authors would like to thank Prof. Y.Ermoliev for his kind advice concerning the convergence of the stochastic approximation method.   They also would like to thank the referees for their kind comments to this paper.   They also want to express their gratitude to Mr. T.Nojima for his kind assistance in preparing the manuscript.

REFERENCES

[1] Y.Y. Haimes, W.A. Hall, and H.F. Friedmann, Multiobjective Optimization in Water Resources Systems, The Surrogate Worth Trade-off Method, Elsevier Scientific, 1975.

[2] A.M. Geoffrion, J.S. Dyer, and A. Feinberg, "An Interactive Approach for Multi-Criterion Optimization with Application to the Operation of Academic Department", Management Science, Vol.19, No.4, pp.357-368, 1972.

[3] A.P. Wierzbicki, "The Use of Reference Objectives in Multiobjective Optimization. Theoretical Implications and Practical Experiences", WP-79-66, IIASA, 1979.

[4] H. Nakayama and Y. Sawaragi, "Satisficing Trade-off Method for Multiobjective Programming", Interactive Decision Analysis, Edited by A.P. Wierzbicki, pp.113-122, Springer-Verlag, 1984.

[5] K.S. Narendra and M.A.L.Thathachar, "Learning Automata - A Survey", IEEE Trans. Systems, Man, and Cybernetics, Vol. 4, pp.323-334, 1974.

[6] N. Baba, New Topics in Learning Automata Theory and Applications, Lecture Notes in Control and Information Sciences, Springer-Verlag, 1985.

[7] J. Matyas, "Random Optimization", Automation and Remote Control, Vol.28, pp.246-253, 1965.

[8] F.J. Solis and R.J. Wets, "Minimization by Random Search Techniques", Mathematics of Operations Research, Vol.6, pp.19-30, 1981.

[9] N. Baba, "Recent Developments in Learning Automata Theory and Their Applications", Modelling and Simulation Methodology in the Artificial Intelligence Era, Edited by M.S. Elzas et al, North-Holland, 1986.

[10] N. Baba, "ε-optimal Nonlinear Reinforcement Scheme Under a Nonstationary Multi-teacher Environment", IEEE Trans. Systems, Man, and Cybernetics, Vol. 14, No.3, pp. 538-541, 1984.

[11] H. Robbins and S. Monro, "A Stochastic Approximation Method", Ann. Math. Statist. , Vol. 22, pp.400-407, 1951.

[12] M.T. Wasan, Stochastic Approximations, Cambridge University Press, 1969.

[13] H.J. Kushner and D.S. Clark, Stochastic Approximation Methods for Constrained and Unconstrained Systems, Springer-Verlag, 1978.

[14] R.L. Kashyap, "Application of Stochastic Approximation", in Adaptive Learning and Pattern Recognition Systems, J.M. Mendel and K.S. Fu, Editors, Academic Press, 1970.

[15] Y. Ermoliev, "Stochastic Quasigradient Methods and Their Application to System Optimization", Stochastics, Vol.9, pp.1-36, 1983.

[16] E.A. Nurminskii, "Convergence Conditions of Stochastic Programming Algorithms", Cybernetics, No. 3, pp.464-468, 1973.

[17] A. Dvoretsky, "On Stochastic Approximation", Proc. 3rd Berkeley Symp. on Math. Stat. and Probability I, pp.39-55, 1956.

A Stochastic Programming Model for Optimal Power Dispatch: Stability and Numerical Treatment

Nicole Gröwe and Werner Römisch

Humboldt-Universität zu Berlin

Fachbereich Mathematik

PSF 1297, 1086 Berlin

Abstract

The economic dispatch of electric power with uncertain demand is modeled as stochastic program with simple recourse. We analyze quantitative stability properties of the power dispatch model with respect to the L_1–distance of the marginal distribution functions of the demand vector. These stability results are used to derive asymptotic properties of the model if the (true) marginal distributions are replaced by smooth nonparametric estimates based on the kernel method. Finally, we discuss how smooth estimates can be used efficiently for the numerical treatment of simple recourse models by using nonlinear programming techniques. Numerical results are reported for Dantzig's Aircraft Allocation Problem.

1 Introduction

A model for the optimal dispatch of electric power to the units of an energy production system is considered that takes explicit account of the uncertainty of the electric power demand. This is done by introducing so-called expected recourse costs for under- and over-dispatching (similar to [5]) and leads to a large-scale convex stochastic program with simple recourse. For the uncertain demand, we suppose that a set of empirical data is given for the whole operating cycle. This motivates investigations in two directions:

(1) the stability analysis of stochastic programs with simple recourse to treat the situation of incomplete information (see Section 3), and

(2) the choice of appropriate estimators for the distribution functions of the random demand at each time interval.

The stability analysis is carried out by applying general results taken from [24], [25]. We also refer to [9] for a survey of stability results in stochastic programming, to [10], [16], [28], [31] for relevant statistical stability results and to [4] where the possibility of using density estimates in stochastic programming is outlined.

For our application, we motivate the use of nonparametric estimators for the distribution functions of the random demand. Nonparametric estimators based on the kernel method (see e.g. [7], [21]) are apparently favourable in our context, since they lead to stochastic programs having the property that the objective function is continuously differentiable and that function (gradient) values can be computed efficiently without numerical integration (cf. Section 5). Additionally, asymptotic properties of kernel-type estimators for distribution functions are comparable to those of the empirical distribution (see Section 4). In Section 4, we also outline how these asymptotic properties together with stability results of Section 3 lead to convergence rates for optimal solution sets of the power dispatch model if the sample size of observations for the demand tends to infinity. In Section 5, we discuss the numerical treatment of simple recourse models involving kernel-type estimators (for the unknown distribution) via standard nonlinear programming techniques. We report on the development of a program system and on numerical results for (a modified version of) Dantzig's Aircraft Allocation Problem.

2 A model for optimal power dispatch with uncertain demand

The problem of optimal power dispatch consists in allocating amounts of electric power to the generation units of an energy production system such that the total generation costs are minimal while the actual power demand is met and certain operational constraints (of the system) are satisfied. The system we shall consider comprises thermal power stations (tps), pumped hydro storage plants (psp) and an energy contract with connected systems.

The peculiarities of the system and the power dispatch model (cf. also [26]) are the following:

(a) tps and psp serve as base– and peak–load plants, respectively,

(b) the model is designed for a daily operating cycle and assumes that a unit commitment stage has been carried out before,

(c) the reserve levels and transmission losses are modeled by means of adjusted portions of the demand,

(d) the cost functions of the thermal plants are taken to be strictly convex and quadratic,

(e) the model takes explicit account of the uncertainty of the electric power demand by introducing an expected recourse action which is associated with the mismatch between scheduled generation and actual demand.

To give a more detailed description of the model, let K and M denote the number of tps and psp, respectively. Assume that the scheduling time–horizon consists of N intervals T_r ($r = 1, \ldots, N$). Let $I_r \subset \{1, 2, \ldots, K\}$ denote the index set of available online tps within the time interval T_r ($r = 1, \ldots, N$). The (unknown) outputs of the tps and psp at the interval T_r are y_{lr} ($l = 1, \ldots, K$) and s_{jr} (generation mode of the psp $j\epsilon\{1, \ldots, M\}$), respectively. By w_{jr} we denote the input of the psp $j\epsilon\{1, \ldots, M\}$ during the pumping mode and by e_r the level of electric power which corresponds to the energy contract at time interval T_r. Denoting $x := (y, s, w, e)^T \epsilon R^m$ with $m := N(K + 2M + 1)$ the model of optimal power dispatch reads

$$\min \{g(x) : x \epsilon C, Ax = z\} \qquad (2.1)$$

Here, the total generation cost function g is convex quadratic on R^m and has the form

$$g(x) := \sum_{r=1}^{N} \left(\sum_{l \epsilon I_r} C_{lr}(y_{lr}) + d_r e_r \right)$$

where $C_{lr}(\cdot)$ are strictly convex quadratic cost functions for the tps l within T_r and d_r is the cost according to the contract at T_r ($l = 1, \ldots, K; r = 1, \ldots, N$).

The set $C \subset R^m$ in (2.1) is a nonempty bounded convex polyhedron formed by the operational constraints of the system, e.g. bounds for the power output of the plants, balances between generation and pumping in the psp, balances for the psp over the whole time–horizon, fuel quotas of the tps.

The equation $Ax = z$ in (2.1) reads componentwise ($r = 1, \ldots, N$)

$$[Ax]_r := \sum_{l \epsilon I_r} y_{lr} + \sum_{j=1}^{M} (s_{jr} - w_{jr}) + e_r = z_r$$

and says that the total generated output meets the demand $z = (z_1, \ldots, z_N)^T$ at each time interval T_r. We consider the demand z as a random vector and denote by μ its probability distribution on R^N and by F_r the distribution function of z_r ($r = 1, \ldots, N$). Following [5] we introduce a penalty cost for the deviation of the scheduled output from the actual demand for under- and over-dispatching, respectively. To be more precise, we define

$$\tilde{Q}(t) := \sum_{r=1}^{N} \tilde{Q}_r(t_r) := \sum_{r=1}^{N} \begin{cases} q_r^+ t_r, & t_r \geq 0 \\ -q_r^- t_r, & t_r < 0 \end{cases} \qquad (t \epsilon R^N) \qquad (2.2)$$

where q_r^+ and q_r^- are the recourse costs for the under- and over-dispatching at time interval T_r ($r = 1, \ldots, N$), respectively. For a

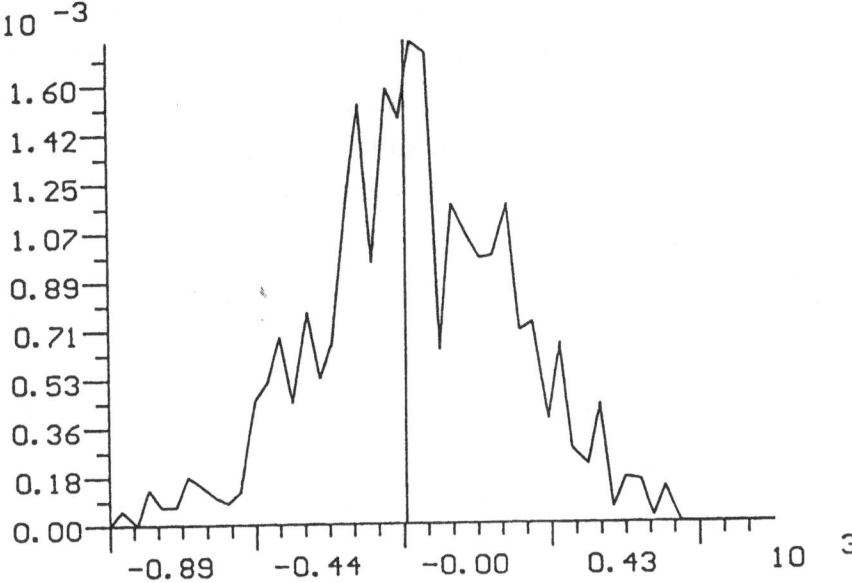

Fig. 2.1

discussion of the interpretation and choice of the recourse costs q_r^+ and q_r^- we refer to [5, pp. 181, 184].

Adding the expected recourse costs $E\left[\tilde{Q}(z - Ax)\right]$ to the deterministic cost function g we arrive at the following stochastic power dispatch model :

$$\min\left\{g(x) + \sum_{r=1}^{N} \int_{R} \tilde{Q}_r(t - [Ax]_r)\, dF_r(t) : x\epsilon C\right\} \qquad (2.3)$$

Similar power dispatch models are considered in [3] and [5]. More information on various aspects of power dispatch can be found in the volume [32] and in several papers of [11], Part IV. It is well-known that (2.3) is a particular stochastic program with simple recourse (see [13]). (2.3) is a large scale convex nonlinear program having linear constraints. If all distribution functions F_r ($r = 1, \ldots, N$) have densities, the objective function of (2.3) is continuously differentiable (cf. [13], p.56 ff.). To give an idea how large (2.3) is for real-life applications, we mention that for the energy production system of East Germany typically $K := 26$, $M := 5$ which leads together with (say)

$N := 24$ to $m := 888$. For the uncertain demand, a set of empirical data (in practice, a medium - sized sample) is given (see also Section 4). It is suggested in [3] and [5] that the distribution of the random demand (at each time interval) can be chosen as (trimmed) normal. However, our tests with the available empirical data did not justify this hypothesis (especially for all time-intervals in the day-time). As an example, Figure 2.1 shows an estimate for the density of the centered demand during the hour 1 p.m. – 2 p.m. (of a day of normal category). The estimate is obtained according to formula (4.1) and by using the triangular kernel (with $b_n = 30$ and $n = 436$). Finally, we preferred the use of nonparametric estimators for the (unknown) distribution functions. This is described in more detail in Section 4 and 5.

3 Stability analysis

Consider the following (convex) stochastic program with simple recourse and random right–hand side

$$\min \{g(x) + Q_\mu(Ax) : x \epsilon C\} \tag{3.1}$$

where

$$\begin{aligned} Q_\mu(\chi) &:= \int_{R^s} \tilde{Q}(z - \chi)\, \mu(dz) \\ \tilde{Q}(t) &:= \min \{q^T y : (I, -I)y = t, y \in R_+^{2s}\} \end{aligned} \tag{3.2}$$

We assume that g is a real-valued convex function on R^m, C is a nonempty, closed, convex subset of R^m, A is a $s \times m$ matrix, $q \epsilon R^{2s}$ and μ is a (Borel) probability measure on R^s.

Under the basic assumptions

(A1) $\qquad\qquad q^+ + q^- \epsilon R_+^s$, where $q = \begin{pmatrix} q^+ \\ q^- \end{pmatrix}$

(A2) $\displaystyle\int_{R^s} \|z\| \, \mu(dz) < +\infty$ ($\|\cdot\|$ denoting the Euclidean norm on R^s),

$\tilde{Q}$ has the representation (2.2) and Q_μ is a real-valued separable convex function on R^s having the shape

$$Q_\mu(\chi) = \sum_{r=1}^{s} \left[q_r^+ (\bar{\mu}_r - \chi_r) - (q_r^+ + q_r^-) \int_{-\infty}^{\chi_r} (t - \chi_r) \, dF_r(t) \right] \quad (3.3)$$

where F_r $(r = 1, \ldots, s)$ are the one–dimensional marginal distribution functions of μ and $\bar{\mu}_r$ $(r = 1, \ldots, s)$ their mean values (cf. e.g.[13]).

This section deals with the stability of problem (3.1), when the underlying probability measure μ is subjected to (small) perturbations. Here, stability means that the optimal value $\varphi(\mu)$ and the optimal solution set $\psi(\mu)$ of problem (3.1) enjoy certain continuity properties with respect to variations of μ in a (properly selected) subset of probability measures on R^s equipped with a suitable distance ("probability metric").

To select a set of probability measures and a metric, we notice that, due to the separability structure of Q_μ (see (3.3)), problem (3.1) only depends on the marginal distributions μ_r $(r = 1, \ldots, s)$ of μ. Hence, we may assume that μ has independent one-dimensional marginal distributions.

Therefore we consider the following metric space $(\mathcal{M}(\mathrm{R}^s), d)$ where $\mathcal{M}(\mathrm{R}^s) := \{\nu : \nu$ is a probability measure on R^s having independent one-dimensional marginals ν_r $(r = 1, \ldots, s)$ and $\int_{\mathrm{R}} |t| \, \nu_r(dt) < \infty$ $(r = 1, \ldots, s)\}$ and

$$d(\nu_1, \nu_2) := \sum_{r=1}^{s} \int_{\mathrm{R}} |F_{1r}(t) - F_{2r}(t)| \, dt, \quad (3.4)$$

F_{1r} and F_{2r} $(r = 1, \ldots, s)$ denoting the one-dimensional marginal distribution functions of $\nu_1, \nu_2 \in \mathcal{M}(\mathrm{R}^s)$.

The first stability result asserts upper semicontinuity of the optimal set mapping $\psi(\cdot)$ and a local Lipschitz property of the optimal value function $\varphi(\cdot)$ of (3.1) at $\mu \in \mathcal{M}(\mathrm{R}^s)$.

Theorem 3.1

Fix $\mu \in \mathcal{M}(\mathbf{R}^s)$, suppose (A1) and let $\psi(\mu)$ be nonempty and bounded.

Then there exist constants $L > 0$, $\delta > 0$ such that

$$|\varphi(\mu) - \varphi(\nu)| \leq L\, d(\mu, \nu)$$

and $\psi(\nu)$ is nonempty whenever $\nu \in \mathcal{M}(\mathbf{R}^s)$, $d(\mu, \nu) < \delta$. The set-valued mapping $\psi(\cdot)$ from $(\mathcal{M}(\mathbf{R}^s), d)$ into $\mathbf{R}^m$ is upper semicontinuous at μ, i.e. for each open set $\mathcal{U}$ containing $\psi(\mu)$ there exists $\delta_0 > 0$ such that $\psi(\nu) \subset \mathcal{U}$ whenever $\nu \in \mathcal{M}(\mathbf{R}^s)$, $d(\mu, \nu) < \delta_0$.

Proof

We apply Theorem 2.4 and Remark 2.5 of [25] and obtain the assertion by using the Wasserstein metric W_1 (cf. Section 2 of [25]) instead of the metric d. It remains to notice that W_1 coincides with d on $\mathcal{M}(\mathbf{R}^s)$ if $\mathbf{R}^s$ is equipped with the norm $\|z\|_1 := \sum_{r=1}^s |z_i|$ $(z \in \mathbf{R}^s)$ (see Remark 2.11 in [25]).●

The following example shows that, under the assumptions of the above Theorem, ψ is in general not lower semicontinuous at μ even if μ has a density. Recall that lower semicontinuity of ψ at $\mu \in \mathcal{M}(\mathbf{R}^s)$ means that for each open set $\mathcal{U}$ satisfying $\mathcal{U} \cap \psi(\mu) \neq \emptyset$ there exists $\delta_0 > 0$ such that $\mathcal{U} \cap \psi(\nu) \neq \emptyset$ whenever $d(\mu, \nu) < \delta_0$.

Example 3.2

In (3.1), let $m = s = 1$, $g(x) \equiv 0$, $C := \mathbf{R}$, $q := (1, 1)^{\mathrm{T}}$, $A := 1$. Consider the family ν_ε, $\varepsilon \in [0, 1]$ of probability measures on $\mathbf{R}$ given by their densities

$$\Theta_\varepsilon(t) := \begin{cases} 1 - \varepsilon & , t \in \left[-1, -\tfrac{1}{2}\right] \\ \varepsilon & , t \in \left(-\tfrac{1}{2}, \tfrac{1}{2}\right) \\ 1 - \varepsilon & , t \in \left[\tfrac{1}{2}, 1\right] \\ 0 & , \text{otherwise.} \end{cases}$$

Then $\tilde{Q}(t) = |t|$ $(t \in \mathbf{R})$, (A1) is satisfied and $\nu_\varepsilon \in \mathcal{M}(\mathbf{R}^s)$ for all $\varepsilon \in [0, 1]$. We obtain from (3.3) that

$$\psi(\nu_\varepsilon) = \{0\} \qquad \text{for all } \varepsilon \in (0,1],$$
$$\psi(\nu_0) = [-\frac{1}{2}, \frac{1}{2}] \qquad \text{and}$$
$$\varphi(\nu_\varepsilon) = \frac{3}{4} - \frac{\varepsilon}{2} \qquad \text{for } \varepsilon \in [0,1].$$

Furthermore, we have $d(\nu_\varepsilon, \nu_0) \leq \varepsilon$. Hence, we conclude that ψ is not lower semicontinuous at $\mu := \nu_0$.

Under a certain positivity condition for the one–dimensional marginal densities of μ, we now show that (at least) the sets of optimal tenders behave locally Hölder continuous at μ.

Theorem 3.3

Fix $\mu \in \mathcal{M}(\mathbf{R}^s)$, suppose $q_r^+ + q_r^- > 0$ $(r = 1, \ldots, s)$ and let $\psi(\mu)$ be nonempty and bounded. Assume, in addition, that the one–dimensional marginal densities $\Theta_r (r = 1, \ldots, s)$ of μ exist and that there exist real numbers $a_r, b_r, \varepsilon > 0$ $(r = 1, \ldots, s)$ such that the conditions $A(\psi(\mu)) \subseteq \times_{r=1}^s (a_r, b_r)$ and $\Theta_r(t) \geq \varepsilon_r$ for all $t \in (a_r, b_r)$ and $r = 1, \ldots, s$ hold.

Then the set $\{Ax : x \in \psi(\mu)\}$ is a singleton and there exist constants $L_1 > 0$ and $\delta_1 > 0$ such that for all $\nu \in \mathcal{M}(\mathbf{R}^s)$ with $d(\mu, \nu) < \delta_1$ we have

$$\sup_{x \in \psi(\nu)} \|Ax - \chi_*\| \leq L_1 d(\mu, \nu)^{1/2}$$

where

$$\{\chi_*\} = \{Ax : x \in \psi(\mu)\}$$

Proof

We want to apply Theorem 4.3 in [24]. To this end we have to show that Q_μ is strongly convex on $V := \times_{r=1}^s (a_r, b_r)$. Let $\lambda \in [0,1]$ and $\chi, \tilde{\chi}$ be chosen such that $\chi_r, \tilde{\chi}_r \in (a_r, b_r)$ for all $r = 1, \ldots, s$. Then we obtain from (3.3)

$$Q_\mu(\lambda\chi + (1-\lambda)\tilde{\chi}) = \lambda Q_\mu(\chi) + (1-\lambda)Q_\mu(\chi) - G(\chi, \tilde{\chi}; \lambda),$$

where

$$G(\chi, \tilde{\chi}; \lambda) := \sum_{r=1}^{s} (q_r^+ + q_r^-)\{\lambda[h_r(\lambda\chi_r + (1-\lambda)\tilde{\chi}_r; \chi_r) - h_r(\chi_r; \chi_r)] + (1-\lambda)[h_r(\lambda\chi_r + (1-\lambda)\tilde{\chi}_r; \tilde{\chi}_r) - h_r(\tilde{\chi}_r; \tilde{\chi}_r)]\}$$

and $h_r(u, v) := \int_{-\infty}^{u}(t-v)\Theta_r(t)\,dt$, $u, v \in \mathbb{R}$, $r = 1, \ldots, s$. Now, let $r \in \{1, \ldots, s\}$ and assume without loss of generality that $\chi_r < \tilde{\chi}_r$. Then we have, setting $\chi_r(\lambda) := \lambda\chi_r + (1-\lambda)\tilde{\chi}_r$,

$$
\begin{aligned}
h_r(\chi_r(\lambda); \chi_r) - h_r(\chi_r; \chi_r) &= \int_{\chi_r}^{\chi_r(\lambda)}(t-\chi_r)\Theta_r(t)\,dt \\
&\geq \varepsilon_r \int_{\chi_r}^{\chi_r(\lambda)}(t-\chi_r)\,dt \\
&= \frac{\varepsilon_r}{2}(\chi_r(\lambda) - \chi_r)^2 \\
&= \frac{\varepsilon_r}{2}(1-\lambda)^2(\chi_r - \tilde{\chi}_r)^2.
\end{aligned}
$$

Analogously, we get the inequality

$$h_r(\chi_r(\lambda); \tilde{\chi}_r) - h_r(\tilde{\chi}_r; \tilde{\chi}_r) \geq \frac{\varepsilon_r}{2}\lambda^2(\chi_r - \tilde{\chi}_r)^2.$$

Altogether, we obtain

$$
\begin{aligned}
Q_\mu(\lambda\chi + (1-\lambda)\tilde{\chi}) &\leq \lambda Q_\mu(\chi) + (1-\lambda)Q_\mu(\tilde{\chi}) \\
&\quad -\frac{1}{2}\sum_{r=1}^{s}(q_r^+ + q_r^-)\varepsilon_r\lambda(1-\lambda)(\chi_r - \tilde{\chi}_r)^2 \\
&\leq \lambda Q_\mu(\chi) + (1-\lambda)Q_\mu(\tilde{\chi}) \\
&\quad -\frac{\kappa}{2}\lambda(1-\lambda)\|\chi - \tilde{\chi}\|^2
\end{aligned}
$$

where $\kappa := \min_{r=1,\ldots,s}(q_r^+ + q_r^-)\varepsilon_r > 0$ and Q_μ is strongly convex on V. Setting $\lambda = \frac{1}{2}$, this together with the convexity of g implies in particular

$$g(x) + Q_\mu(Ax) \geq \varphi(\mu) + \frac{\kappa}{4}\|Ax - Ax_*\|^2$$

for all $x \in C$ and $x_* \in \psi(\mu)$. This proves that the set $\{Ax : x \in \psi(\mu)\}$ is a singleton. The assertion now follows from Theorem 4.3 in [24] with the same argument concerning the metrics as in the proof of Theorem 3.1. ●

Remark 3.4

Example 4.5 in [24] shows that the exponent $1/2$ on the right-hand side in the assertion of Theorem 3.3 is optimal, and our Example 3.2 shows that the assertion of Theorem 3.3 is not true if $A(\psi(\mu))$ is not contained in the support of μ.

Theorem 3.5

Let, in addition to the assumptions of Theorem 3.3, g be convex quadratic and C be polyhedral.

Then there exist constants $L_2 > 0$ and $\delta_2 > 0$ such that

$$d_H(\psi(\mu), \psi(\nu)) \leq L_2\, d(\mu, \nu)^{1/2}$$

whenever $\nu \in \mathcal{M}(\mathbb{R}^s)$, $d(\mu, \nu) < \delta_2$. (Here, d_H denotes the Hausdorff distance on subsets of $\mathbb{R}^m$.)

Proof

The result follows from Theorem 2.7 in [25] by repeating the strong-convexity and metric arguments in the proof of Theorem 3.3.

●

Remark 3.6

The discussion in Remark 2.9 in [25] shows that Theorem 3.5 does not remain true for a general convex constraint set C and for a general convex (deterministic) objective function g. Fortunately, the above results cover the situation of the power dispatch model in Section 2, if the marginal densities of μ fulfil the positivity condition imposed in Theorem 3.3.

Extensions of our stability results to more general recourse models may be found in [24], [25] and in the papers [14], [23], where qualitative stability results for general recourse problems are obtained with respect to the topology of weak convergence on the set of all probability measures (cf. [2]).

4 Smooth nonparametric distribution estimates and asymptotic analysis

In this section, we consider nonparametric estimates for univariate distribution functions and analyze their rates of convergence. In particular, we study smooth estimates which are obtained by integrating a density estimator of the kernel type. This is motivated by the stochastic power dispatch model (2.3), since there the distribution functions of the uncertain electric power demand at each time interval have to be estimated and since smooth estimates lead to a smooth nonlinear programming problem.

Let $(X_i)_{i \in \mathbb{N}}$ be a sequence of independent and identically distributed real–valued random variables with common distribution function F. By $\mathcal{F}_n$ we denote the empirical distribution function for sample size $n \in \mathbb{N}$, i.e.

$$\mathcal{F}_n(x) := n^{-1} \sum_{i=1}^{n} I_0(x - X_i) \quad (x \in \mathbb{R})$$

where I_0 is the indicator function of the interval $[0, +\infty)$. A nonnegative function k having the property $\int_{\mathbb{R}} k(x)\, dx = 1$ is called *kernel*. Suppose (b_n) is a sequence of positive numbers (*"smoothing parameters"*) tending to zero. Then

$$\hat{f}_n(x) := (nb_n)^{-1} \sum_{i=1}^{n} k\left((x - X_i)b_n^{-1}\right) \quad (x \in \mathbb{R}) \qquad (4.1)$$

is a *kernel estimate for the density* $f := F'$ and the corresponding kernel estimate of F is

$$\hat{\mathcal{F}}_n(x) := \int_{-\infty}^{x} \hat{f}_n(t)\, dt = \int_{-\infty}^{\infty} \mathcal{K}\left((x - t)b_n^{-1}\right)\, d\mathcal{F}_n(t) \qquad (4.2)$$

where $\mathcal{K}(x) := \int_{-\infty}^{x} k(t)\, dt$. $\hat{\mathcal{F}}_n$ may be interpreted as a smoothed version of the empirical distribution function $\mathcal{F}_n$. For more information and background on kernel-type estimators it is referred to [7],[21] and [30].

In the following, a kernel k is called *class s kernel* for some $s \in \mathbb{N}$ if

$$\int_{\mathbb{R}} x^i k(x) \, dx = 0 \quad , i = 1, \ldots, s-1,$$

$$\int_{\mathbb{R}} |x|^s k(x) \, dx < \infty.$$

If, in addition, the kernel k is symmetric (about 0), we need only consider even values of s. In that case, it is known that class 4 kernels of compact support do not exist (see [7, p.100], [30, p.66]). For a discussion of class s kernels which are possibly negative-valued see [7, Chapter 7.2].

Some kernels, which are, in fact, all symmetric class 2 kernels and will be considered in Section 5, and their cumulative distribution functions $\mathcal{K}$ are now listed.

Epanechnikov

$$k(t) = \begin{cases} \frac{3}{4\sqrt{5}}(1 - t^2/5) & (|t| \leq \sqrt{5}) \\ 0 & otherwise \end{cases}$$

$$\mathcal{K}(t) = \begin{cases} 0 & (t \leq -\sqrt{5}) \\ \frac{1}{2} + \frac{3t}{4\sqrt{5}} - \frac{t^3}{20\sqrt{5}} & (-\sqrt{5} < t < \sqrt{5}) \\ 1 & (t \geq \sqrt{5}) \end{cases}$$

Biweight

$$k(t) = \begin{cases} \frac{15}{16}(1 - t^2)^2 & (|t| \leq 1) \\ 0 & otherwise \end{cases}$$

$$\mathcal{K}(t) = \begin{cases} 0 & (t \leq -1) \\ \frac{1}{2} + \frac{15}{16}t - \frac{5}{8}t^3 + \frac{3}{16}t^5 & (-1 < t < 1) \\ 1 & (t \geq 1) \end{cases}$$

Triangular

$$k(t) = \begin{cases} 1 - |t| & (|t| \leq 1) \\ 0 & otherwise \end{cases}$$

$$K(t) = \begin{cases} 0 & (t \leq -1) \\ \frac{1}{2} + t + \frac{1}{2}t^2 & (-1 < t \leq 0) \\ \frac{1}{2} - t + \frac{1}{2}t^2 & (0 < t < 1) \\ 1 & (t \geq 1) \end{cases}$$

Let $C_b^s := C_b^s(\mathrm{R})$ denote the class of s–times continuously differentiable functions F such that $F^{(s)}$ is bounded on R. The following auxiliary result gives an estimate for the distance

$$\|\hat{\mathcal{F}}_n - F\|_\infty := \sup_{x \in \mathrm{R}} |\hat{\mathcal{F}}_n(x) - F(x)|$$

where $\hat{\mathcal{F}}_n$ is the kernel type estimate (4.2) for a sufficiently smooth distribution function F. Its proof follows essentially the ideas developed in [34].

Lemma 4.1

Let $s \in \mathrm{N}$ and assume that $F \in C_b^s$ and k is a class s kernel. Then

$$\|\hat{\mathcal{F}}_n - F\|_\infty \leq Cb_n^s + \|\mathcal{F}_n - F\|_\infty \qquad (n \in \mathrm{N})$$

where

$$C := \frac{1}{s!}\|F^{(s)}\|_\infty \int_{\mathrm{R}} |x|^s k(x)\, dx.$$

Proof

From [34], Lemma 2.3 we have the estimate

$$\|\hat{\mathcal{F}}_n - F\|_\infty \leq \|\mathcal{F}_n - F\|_\infty + \sup_{x \in \mathrm{R}} |\mathrm{E}\,\hat{\mathcal{F}}_n(x) - F(x)|$$

where E denotes the mean value with respect to the sample probability space. It remains to derive an estimate for the second term on the right-hand side. For each $x \in \mathrm{R}$ we obtain by Taylor's expansion and using that k is a class s kernel

$$\begin{aligned} \mathrm{E}\,\hat{\mathcal{F}}_n(x) - F(x) &= \int_{\mathrm{R}} K((x-t)b_n^{-1})\, dF(t) - F(x) \\ &= \int_{\mathrm{R}} [F(x - tb_n) - F(x)]\, k(t)\, dt \\ &= \int_{\mathrm{R}} (-tb_n)^s \frac{F^{(s)}(x - \theta tb_n)}{s!} k(t)\, dt \end{aligned}$$

with some $\theta \in (0,1)$ depending possibly on x, t and n. This finally yields the desired estimate.●

Another, but similar, technique for estimating

$$\|\hat{\mathcal{F}}_n - F\|_\infty$$

can be found in [29], Chapt. 23.2. The next result now asserts rates for the almost sure and mean convergence of $\left(\|\hat{\mathcal{F}}_n - F\|_\infty\right)_{n \in \mathbb{N}}$.

Proposition 4.2

Let $s \in \mathbb{N}$, assume that $F \in C_b^s$ and k is a class s kernel. Suppose that (b_n) is chosen such that

$$\limsup_{n\to\infty} b_n^s n^{\frac{1}{2}} < \infty.$$

(a) The following law of the iterated logarithm (LIL) holds

$$\limsup_{n\to\infty} (2n/\log\log n)^{\frac{1}{2}} \|\hat{\mathcal{F}}_n - F\|_\infty \leq 1$$

almost surely.

(b)

$$\limsup_{n\to\infty} n^{\frac{1}{2}} \mathrm{E}(\|\hat{\mathcal{F}}_n - F\|_\infty < \infty.$$

Proof
Part (a) of the assertion follows from Lemma 4.1 and the Smirnov-Chung LIL for empirical distribution functions (see [12, Chapt.6.8], [29]; cf. also the proof of Theorem 3.2 in [34] which in fact is the particular case of our result for $s = 2$).

To establish (b), we again use Lemma 4.1. It remains to apply the following known result for empirical distribution functions (cf. [12, Chapt.3.3]):

$$\mathrm{E}\left[\|\mathcal{F}_n - F\|_\infty\right] \leq 2\,\mathrm{E}\left[\sup_{t\in\mathbb{R}} [\mathcal{F}_n(t) - F(t)]\right] = O(n^{-\frac{1}{2}}) \quad ●$$

Less seems to be known for the convergence of $(\hat{\mathcal{F}}_n)$ to F in terms of the L_1-distance. Next we give a speed-of-convergence result in this direction for the case that the kernel k has compact support.

Proposition 4.3

In addition to the hypotheses of Proposition 4.2, let the kernel k have compact support and assume that the p-th absolute moment $M_p := \int_{\mathbb{R}} |t|^p \, dF(t)$ be finite for some $p > 1$. Then

$$\limsup_{n \to \infty} n^{\frac{1}{2}(1-\frac{1}{p})} \mathrm{E}\left(\int_{\mathbb{R}} |\hat{\mathcal{F}}_n(t) - F(t)| \, dt\right) < \infty.$$

Proof

Let $\varepsilon_n := n^{-\frac{1}{2}}$, $c_n := \varepsilon_n^{-\frac{1}{p}}$ for each $n \in \mathbb{N}$. Then we obtain

$$
\begin{aligned}
n^{\frac{1}{2}(1-\frac{1}{p})} \mathrm{E}\left(\int_{\mathbb{R}} |\hat{\mathcal{F}}_n(t) - F(t)| \, dt\right) \;\le\; & c_n^{p-1} \mathrm{E}\Big(\int_{-\infty}^{-c_n} (\hat{\mathcal{F}}_n(t) + F(t)) \, dt \\
& + 2c_n \|\hat{\mathcal{F}}_n - F\|_\infty \\
& + \int_{c_n}^{\infty} ((1 - \hat{\mathcal{F}}_n(t)) + (1 - F(t))) \, dt\Big) \\
= \; & c_n^{p-1}\Big[\int_{c_n}^{\infty} (F(-t) + (1 - F(t))) \, dt \\
& + \int_{c_n}^{\infty} (\mathrm{E}\,\hat{\mathcal{F}}_n(-t) + (1 - \mathrm{E}\,\hat{\mathcal{F}}_n(t))) \, dt\Big] \\
& + 2\varepsilon_n^{-1} \mathrm{E}(\|\hat{\mathcal{F}}_n - F\|_\infty).
\end{aligned}
$$

In view of Proposition 4.2(b) we need to study only the first two terms in the last row. By Markov's inequality we have

$$F(-t) + (1 - F(t)) \le 2M_p t^{-p} \quad \text{for all } t > 0.$$

Hence, for the first integral we obtain the estimate

$$c_n^{p-1} \int_{c_n}^{\infty} (F(-t) + (1 - F(t))) \, dt \le c_n^{p-1} 2M_p \int_{c_n}^{\infty} t^{-p} \, dt = \frac{2M_p}{p - 1}.$$

It remains to estimate the second integral. To this end, let I denote the compact support of k and let $n_o \in \mathbb{N}$ be such that $c_n \ge 1$ and $b_n I \subseteq [-\frac{1}{2}, \frac{1}{2}]$ for all $n \ge n_o$.

Let $n \ge n_o$ and $x \ge 1$. We obtain by Markov's inequality

$$
\begin{aligned}
\mathrm{E}\,\hat{\mathcal{F}}_n(-x) &= \int_{\mathbb{R}} F(-x - tb_n) k(t) \, dt = \int_I F(-(x + tb_n)) k(t) \, dt \\
&\le \int_I M_p(x + tb_n)^{-p} k(t) \, dt \le M_p\left(x - \frac{1}{2}\right)^{-p} \le M_p\left(\frac{x}{2}\right)^{-p}.
\end{aligned}
$$

Analogously, we have the estimate

$$1 - \mathrm{E}\,\hat{\mathcal{F}}_n(x) = \int_I (1 - F(x - tb_n))k(t)\,dt \le M_p\left(\frac{x}{2}\right)^{-p}.$$

The last two inequalities yield

$$c_n^{p-1} \int_{c_n}^{\infty} (\mathrm{E}\,\hat{\mathcal{F}}_n(-t) + (1 - \hat{\mathcal{F}}_n(t)))\,dt \le 2^{p+1} M_p c_n^{p-1} \int_{c_n}^{\infty} t^{-p}\,dt = \frac{2^{p+1} M_p}{p-1}.$$

This completes the proof.•

Remark 4.4

It is clear from the proof of Proposition 4.3 that stronger moment conditions for F lead to more comfortable rates of convergence for

$$\mathrm{E}\left(\int_{\mathbb{R}} |\hat{\mathcal{F}}_n(t) - F(t)|\,dt\right) \quad \text{as } n \to \infty.$$

It is not known whether the rate $O\left(n^{-\frac{1}{2}}\right)$ as $n \to \infty$ can be attained as in Proposition 4.2(b).

If F is the uniform distribution on $[0, 1]$ and $\mathcal{F}_n$ the corresponding empirical distribution function, then it is shown in [8, Chapter 6] that there exists a constant $C > 0$ such that

$$C^{-1} n^{-\frac{1}{2}} \le \mathrm{E}\left(\int_{\mathbb{R}} |\mathcal{F}_n(t) - F(t)|\,dt\right) \le C n^{-\frac{1}{2}} \quad \text{for all } n.$$

Remark 4.5

The convergence results show that the speed of convergence of perturbed (or smoothed) empirical distribution function $\hat{\mathcal{F}}_n$ (obtained by the kernel method) to the (unknown) distribution function F is essentially the same as for $\mathcal{F}_n$ if the sequence (b_n) of smoothing parameters is chosen appropriately. If $F \in C_b^s$ and if a class s kernel is used, then $b_n := n^{-\alpha}$ with $\alpha \ge \frac{1}{2s}$ is an appropriate choice. For a thorough discussion of the choice of smoothing parameters (e.g. also for small sample sizes) we refer to [1] and [30] (see also Section

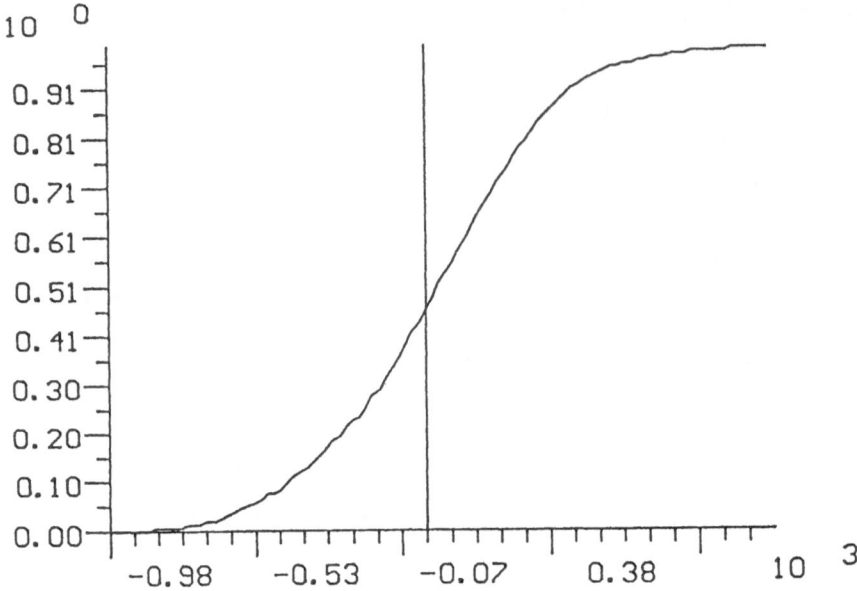

Fig. 4.1

5). Let us consider a stochastic program (3.1) with simple recourse and random right-hand side and suppose that for all components of the random vector a sample of n observations is given. We suppose that all marginal distribution functions are estimated by the kernel method$((4.2))$ which leads to an estimated distribution $\hat{\mu}_n$. Then the stability results of Section 3 together with Propositions 4.2 and 4.3 yield asymptotic properties for the almost sure and mean convergence of the sequences

$$(|\varphi(\mu) - \varphi(\hat{\mu}_n)|)_{n \in N},$$
$$(d_H(\psi(\mu), \psi(\hat{\mu}_n)))_{n \in N}.$$

For example, convergence rates for $\mathrm{E}(d_H(\psi(\mu), \psi(\hat{\mu}_n)))$ $(n \in N)$ can be obtained following the ideas of Corollary 2.12 in [25]. For the power dispatch model of Section 2 we get, in particular,

$$\mathrm{E}\left(d_H(\psi(\mu), \psi(\hat{\mu}_n))\right) = O\left(n^{-1/4}\right),$$

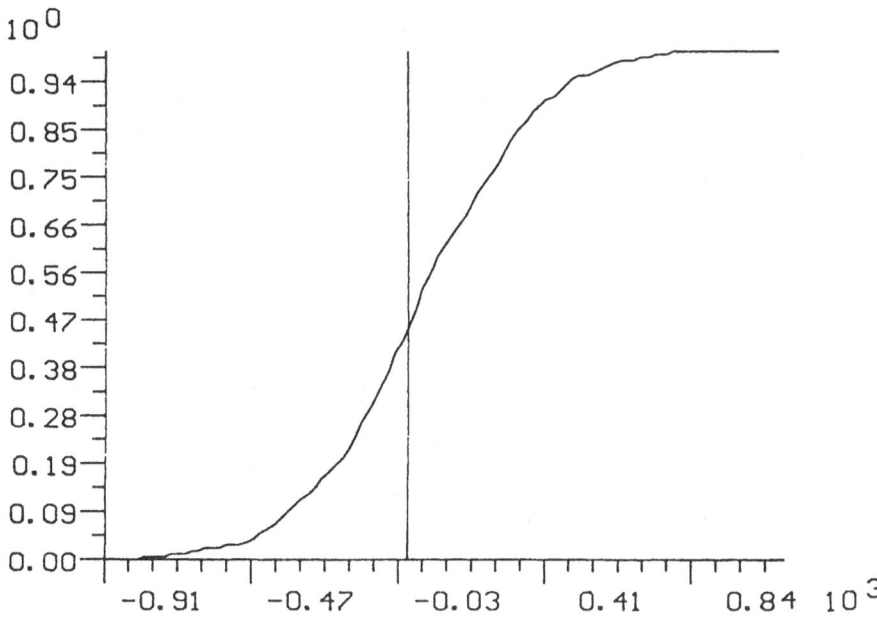

Fig. 4.2

since the support of the random demand is compact and the positivity condition for the marginal densities (see Theorem 3.3) in a neighbourhood of the optimal tender is (certainly) satisfied. This asymptotic argument also provides a certain justification for our numerical approach to the treatment of the economic dispatch in the energy production system in East Germany. The available empirical data is the difference of actual and predicted demand (so-called residuum) over several years. The data has been analyzed and classified (in particular, into data belonging to days with comparable demand curves) (for details see [20]). After performing the data analysis and adding the predicted demand, the sample (for the demand) may be viewed as independent. The following pictures show kernel estimates for the marginal distribution function of the demand during the hour 1 p.m.- 2 p.m. (of a day of normal category). In both cases the triangular kernel has been used with a sample size $n := 436$ and smoothing parameters $b_n = 30$ (Fig. 4.1) and $b_n = 50$ (Fig. 4.2).

5 Numerical treatment, implementation and test example

In this section, we deal with the numerical treatment of stochastic programs with simple recourse and random right–hand side

$$\min \{g(x) + Q_\mu(\chi) : x \epsilon C, Ax = \chi\} \tag{5.1}$$

where g is convex and continuously differentiable, $C \subset \mathbb{R}^m$ is a convex polyhedron, A an $s \times m$ – matrix, Q_μ is defined by (3.2) and has the special feature that only a sample of n observations (with common distribution μ) is available. Since Q_μ only depends on the marginal distribution functions F_r, $r = 1, \ldots, s$ (cf. (3.3)), we may assume that n real observations $X_{r1}, \ldots, X_{rn}$ with common distribution function F_r ($r = 1, \ldots, s$) are given. Our approach is the following: For each $r = 1, \ldots, s$, F_r is estimated by a smooth nonparametric estimator $\hat{F}_r$ based on the kernel k (see Section 4) and F_r in (3.3) is replaced by $\hat{F}_r$. This leads to the following (convex) nonlinear program having continuously differentiable objective and linear constraints:

$$\min \left\{ g(x) + \hat{Q}_\mu(\chi) : x \epsilon C, Ax = \chi \right\} \tag{5.2}$$

$$\hat{Q}(\chi) := \sum_{r=1}^{s} [q_r^+(\hat{\mu}_r - \chi_r) - (q_r^+ + q_r^-)\frac{1}{n} \sum_{i=1}^{n} \{(X_{ri} - \chi_r)\, \mathcal{K}_1((\chi_r - X_{ri})b^{-1})$$
$$+ b\, \mathcal{K}_2((\chi_r - X_{ri})b^{-1})\}]$$

where $\hat{\mu}_r := \frac{1}{n}\sum_{i=1}^{n} X_{ri}$, $\mathcal{K}_1(u) := \int_{-\infty}^{u} k(t)\, dt$, $\mathcal{K}_2(u) := \int_{-\infty}^{u} tk(t)\, dt$ and b is the smoothing parameter.

We note that for many kernels k (especially those mentioned in Section 4) the functions $\mathcal{K}_1$ and $\mathcal{K}_2$ can be computed explicitly. Hence, no numerical integration has to be performed when $\hat{Q}$ and their partial derivatives

$$\frac{\partial \hat{Q}}{\partial \chi_r} := -q_r^+ + (q_r^+ + q_r^-)\frac{1}{n} \sum_{i=1}^{n} \mathcal{K}_1((\chi_r - X_{ri})b^{-1})$$

are evaluated. Hence, the nonlinear program (5.2) can be solved efficiently by using standard nonlinear programming systems, like e.g. the MINOS–system (see [17]).

The essential difference of our approach to those developed in several papers (e.g. [3], [5], [18], [19], [22], [33]) is that we do not consider F_r as discrete distribution functions but replace F_r by a smooth estimate $\hat{F}_r$ ($r = 1, \ldots, s$). This is certainly justified if (at least) medium – sized samples for F_r are given (as in our applications to power dispatch, see Section 2). We also refer to [11] as general reference for numerical methods in stochastic programming and to [18] and [27] for the description of program systems designed for simple recourse problems.

A program system STOCHOPT according to the above mentioned approach has been developed for IBM/PC AT computers. The programs are written in FORTRAN 77 and Turbo Pascal (user interface).

The system consists of the following main parts:

(i) User interface: it realizes an interactive facility for problem specification, the construction of samples (if not available a priorily), choice of smoothing parameters, graphical representation of the optimal solution.

(ii) Nonlinear programming part for solving (5.2).

In (ii) the alternative use of Epanechnikov, biweigth or triangular kernels is possible. Since these kernels have compact support, an appropriate ordering of the samples accelerates the evaluation of $\hat{Q}$ and its gradient considerably. The nonlinear programming part of STOCHOPT has been tested successfully on both modifications of problems taken from [15] and the model for optimal power dispatch. The latter model has been solved for a real-life situation (with $m = 888$). Numerical results will be reported elsewhere. Now, we report the solution of the classical Aircraft Allocation Problem due to Dantzig ([6]).

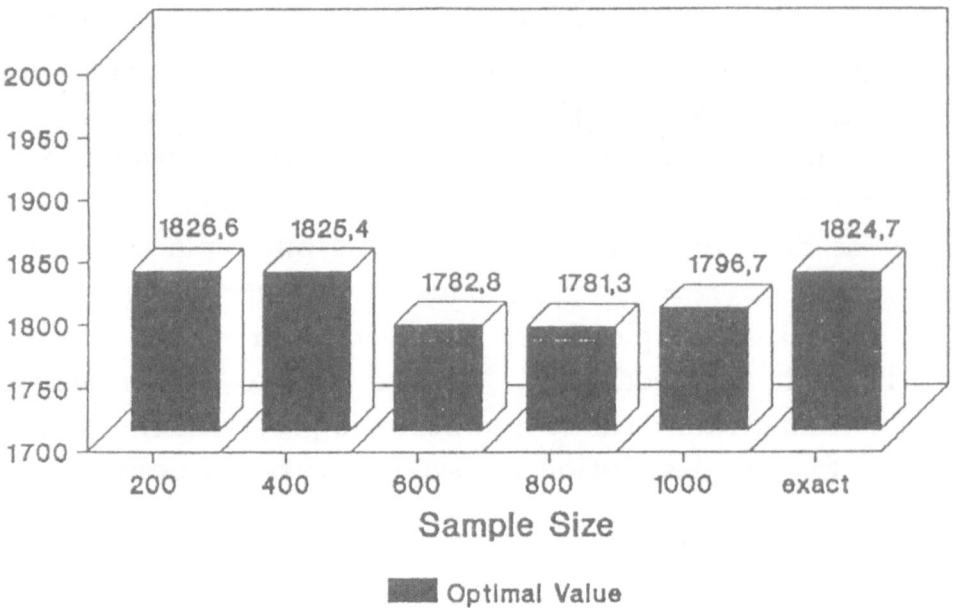

Fig. 5.1 Dependence of the optimal value on the sample size
(with smoothing parameter b_{opt})

An airline wishes to allocate airplanes of various types among its routes to satisfy an uncertain passenger demand, while operating costs plus the lost revenue from passengers turned away are minimal.

The stochastic program with simple recourse is as follows

$$\min \left\{ d^T x + Q(\chi) : Tx = e, Ax = \chi \right\} \qquad (5.3)$$

where $x \in R_+^{17}$, $\chi \in R^5$, $A \in L\ (R^{17}, R^5)$, $T \in L\ (R^{17}, R^4)$ and

$d \in R^{17}, e \in R^4$:

$$A = \begin{pmatrix} 16 & 0 & 0 & 0 & 0 & 0 & 0 & 0 & 0 & 0 & 0 & 0 & 9 & 0 & 0 & 0 & 0 \\ 0 & 15 & 0 & 0 & 0 & 10 & 0 & 0 & 0 & 5 & 0 & 0 & 0 & 11 & 0 & 0 & 0 \\ 0 & 0 & 28 & 0 & 0 & 0 & 14 & 0 & 0 & 0 & 0 & 0 & 0 & 0 & 22 & 0 & 0 \\ 0 & 0 & 0 & 23 & 0 & 0 & 0 & 15 & 0 & 0 & 7 & 0 & 0 & 0 & 0 & 17 & 0 \\ 0 & 0 & 0 & 0 & 81 & 0 & 0 & 0 & 57 & 0 & 0 & 29 & 0 & 0 & 0 & 0 & 55 \end{pmatrix}$$

$$T = \begin{pmatrix} 1 & 1 & 1 & 1 & 1 & 0 & 0 & 0 & 0 & 0 & 0 & 0 & 0 & 0 & 0 & 0 & 0 \\ 0 & 0 & 0 & 0 & 0 & 1 & 1 & 1 & 1 & 0 & 0 & 0 & 0 & 0 & 0 & 0 & 0 \\ 0 & 0 & 0 & 0 & 0 & 0 & 0 & 0 & 0 & 1 & 1 & 1 & 0 & 0 & 0 & 0 & 0 \\ 0 & 0 & 0 & 0 & 0 & 0 & 0 & 0 & 0 & 0 & 0 & 0 & 1 & 1 & 1 & 1 & 1 \end{pmatrix}$$

$$d = (18 \quad 21 \quad 18 \quad 16 \quad 10 \quad 15 \quad 16 \quad 14 \quad 9 \quad 10 \quad 9 \quad 6 \quad 17 \quad 16 \quad 17 \quad 15 \quad 10)^T$$

$$e = (10 \quad 19 \quad 25 \quad 15)^T.$$

The revenue lost per passengers turned away on the r-th route $(r = 1, \ldots, 5)$ is

$$q^+ = (13 \quad 13 \quad 7 \quad 7 \quad 1)^T$$
$$q^- = (0 \quad 0 \quad 0 \quad 0 \quad 0)^T$$

Dantzig solved the program (5.3) with discretely distributed passenger demand z for each of the routes. We assume the passenger demand to have a continuous distribution:

$$\begin{aligned} z_1 &\sim U[200, 300] \\ z_2 &\sim U[50, 150] \\ z_3 &\sim U[140, 220] \\ z_4 &\sim U[10, 340] \\ z_5 &\sim U[580, 620] \end{aligned} \tag{5.4}$$

($U[a, b]$ denotes the uniform distribution on [a,b]). The optimal solution set is listed below:

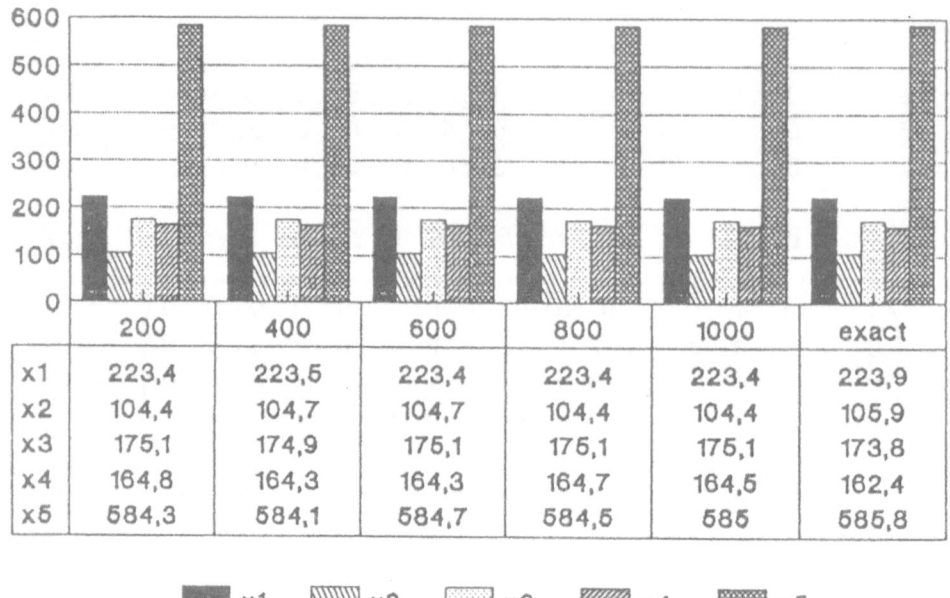

	200	400	600	800	1000	exact
x1	223,4	223,5	223,4	223,4	223,4	223,9
x2	104,4	104,7	104,7	104,4	104,4	105,9
x3	175,1	174,9	175,1	175,1	175,1	173,8
x4	164,8	164,3	164,3	164,7	164,5	162,4
x5	584,3	584,1	584,7	584,5	585	585,8

■ x1   ▨ x2   ▢ x3   ▨ x4   ▨ x5

Fig. 5.2 Optimal tenders for various sample sizes
(with smoothing parameter b_{opt}) and the exact result

Aircrafttype→	1	2	3	4	Tenders
↓ Route					
1	$x_1 = 10$			$x_{13} = 7.1$	$\chi_1 = 224$
2	$x_2 = 0$	$x_6 = 8.2$	$x_{10} = 4.8$	$x_{14} = 0$	$\chi_2 = 106$
3	$x_3 = 0$	$x_7 = 0$		$x_{15} = 7.9$	$\chi_3 = 174$
4	$x_4 = 0$	$x_8 = 10.8$	$x_{11} = 0$	$x_{16} = 0$	$\chi_4 = 162$
5	$x_5 = 0$	$x_9 = 0$	$x_{12} = 20.2$	$x_{17} = 0$	$\chi_5 = 586$

The optimal value of (5.3) is 1824.7.

In order to test our numerical approach, a pseudo-random number generator has been used to simulate samples from the distributions (5.4). The problem (5.2) (as approximation for (5.3)) has been
solved for different sample sizes n and smoothing parameters b, re-

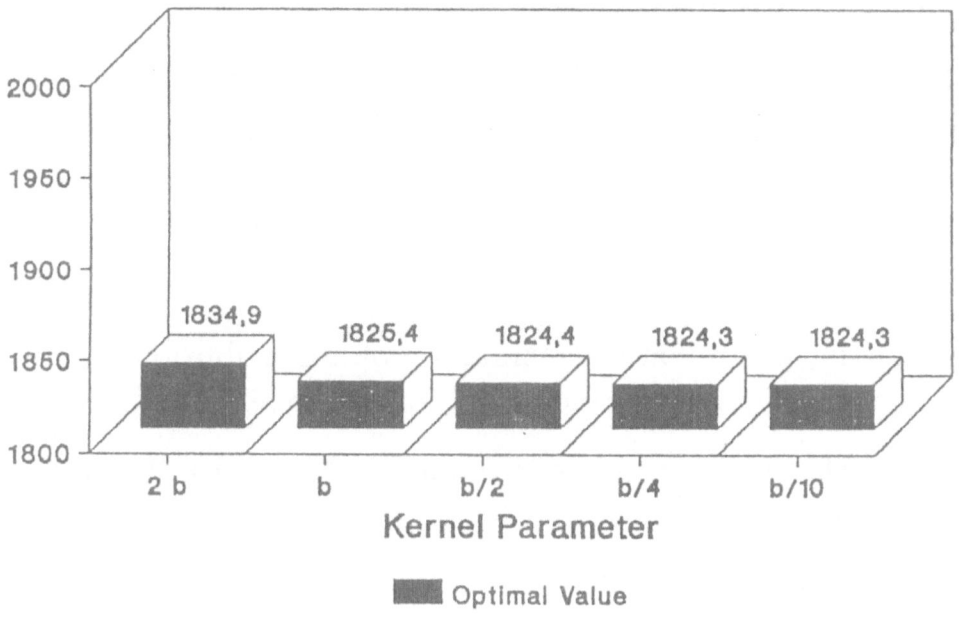

Fig. 5.3 Dependence of the optimal value on the choice of
smoothing parameter $(b = b_{opt})$ for a fixed sample
of 400 observations

spectively. According to the suggestion in [1], the special smoothing
parameter $b_{opt} := 0.5\sigma n^{-\frac{1}{3}}$ (σ being the standard deviation of the unknown distribution) was used when varying the sample size n. We
note that for this choice of b_n the asymptotic properties of Section
4 are valid (see Remark 4.5). The numerical results are summarized
in the figures 5.1-5.3.

The results show that even for small sample sizes good approximations are obtained for optimal values and optimal tenders, respectively. Fig. 5.3 indicates that the optimal value behaves insensitive
on the choice of smoothing parameters (of a relatively large band–
width), but that a choice of smaller smoothing parameters (than that
suggested in [1]) might be favourable.

Acknowledgement

We wish to thank Prof. Georg Pflug (University of Vienna) for beneficial discussions during the GAMM/IFIP–Workshop at Munich. We extend our gratitude to the referees for their constructive criticism.

References

[1] A. Azzalini, A note on the estimation of a distribution function and quantiles by a kernel method, Biometrika 68(1981), 326-328.

[2] P. Billingsley, Convergence of Probability Measures, Wiley, New York, 1968.

[3] J. Böttcher, Stochastische lineare Programme mit Kompensation, Mathematical Systems in Economics, vol. 115, Athenäum Verlag, Frankfurt am Main, 1989.

[4] C. Bouza and S. Allende, Density function estimation and the approximation of convergence rates in stochastic linear programming, Revista de Investigacion Operational 10(1989), 135-140.

[5] D.W. Bunn and S.N. Paschentis, Development of a stochastic model for the economic dispatch of electric power, European Journal of Operational Research 27(1986), 179-191.

[6] G. Dantzig, Linear Programming and Extensions, Princeton University Press, 1963.

[7] L. Devroye, A Course in Density Estimation, Birkhäuser, Boston, 1987.

[8] R.M. Dudley, The speed of the mean Glivenko-Cantelli-convergence, The Annals of Mathematical Statistics 40(1969), 40-50.

[9] J. Dupačová, Stability and sensitivity analysis for stochastic programming, Annals of Operations Research 27(1991), 115-142.

[10] J. Dupačová and R.J.-B. Wets, Asymptotic behaviour of statistical estimators and of optimal solutions of stochastic optimization problems, The Annals of Statistics 16(1988), 1517-1549.

[11] Yu. Ermoliev and R.J.-B. Wets (Eds.), Numerical Techniques for Stochastic Optimization, Springer-Verlag, Berlin, 1988.

[12] P. Gänssler and W. Stute, Wahrscheinlichkeitstheorie, Springer-Verlag, 1977.

[13] P. Kall, Stochastic Linear Programming, Springer-Verlag, Berlin, 1976.

[14] P. Kall, On approximations and stability in stochastic programming, Parametric Optimization and Related Topics (J. Guddat, H.Th. Jongen, B. Kummer, F. Nožička, Eds.), Akademie-Verlag, Berlin, 1987, 387-407.

[15] A.J. King, Stochastic Programming Problems: Examples from the literature, in [11], 543-567.

[16] A.J. King and R.J.-B. Wets, Epi-consistency of convex stochastic programs, Research Report, IBM Research Division, T.J. Watson Research Center, New York, 1989.

[17] B. Murtagh and M. Saunders, Large-scale linearly constrained optimization, Mathematical Programming 14(1978), 41-72.

[18] L. Nazareth, Design and implementation of a stochastic programming optimizer with recourse and tenders, in [11], 273-294.

[19] L. Nazareth and R.J.-B. Wets, Nonlinear Programming Techniques applied to stochastic programs with recourse, in [11], 95-121.

[20] J. Polzehl and K. May, Short term prediction of electric load in a power network, Manuscript, Fachbereich Mathematik der Humboldt-Universität zu Berlin (in preparation).

[21] B.L.S. Prakasa Rao, Nonparametric Functional Estimation, Academic Press, New York, 1983.

[22] A. Prékopa, Dual method for the solution of a one-stage stochastic programming problem with random rhs obeying a discrete probability distribution, Zeitschrift für Operations Research 34(1990), 441-461.

[23] S.M. Robinson and R.J.-B. Wets, Stability in two-stage stochastic programming, SIAM Journal on Control and Optimization 25(1987), 1409-1416.

[24] W. Römisch and R. Schultz, Stability of solutions for stochastic programs with complete recourse having $C^{1,1}$ data, Manuskript, Institut für Operations Research der Universität Zürich, 1989.

[25] W. Römisch and R. Schultz, Stability analysis of stochastic programs, Annals of Operations Research 29(1991) (to appear).

[26] W. Römisch and R. Schultz, Stochastic programs with complete recourse: Stability and an application to power dispatch, in: System Modelling and Optimization (H.-J. Sebastian, K. Tammer,Eds.), Proceedings 14th IFIP Conference (Leipzig, 1989), Lecture Notes in Control and Information Sciences vol.143, Springer-Verlag, Berlin, 1990.

[27] N. Roenko, V. Loskutov and S. Uryasyev, Stochastic nonlinear programming system, IIASA, Laxenburg, Working Paper WP-89-075(1989).

[28] A. Shapiro, Asymptotic properties of statistical estimators in stochastic programming, The Annals of Statistics 17(1989), 841-858.

[29] G.R. Shorack and J. Wellner, Empirical Processes with Applications to Statistics, Wiley, New York, 1986.

[30] B.W. Silvermann, Density Estimation for Statistics and Data Analysis, Chapman and Hall, London, 1986.

[31] S. Vogel, Stability results for stochastic programming problems, Optimization 19(1988), 269-288.

[32] Hj. Wacker (Ed.), Applied Optimization Techniques in Energy Problems, Teubner, Stuttgart, 1985.

[33] R.Wets, Solving stochastic programs with simple recourse, Stochastics 10(1983), 219-242.

[34] B.B. Winter, Convergence rate of perturbed empirical distribution functions, Journal of Applied Probability 16(1979), 163-173.

COMPUTATIONAL TECHNIQUES FOR PROBABILISTIC CONSTRAINED OPTIMIZATION PROBLEMS

János Mayer*
Institute for Operations Research, University of Zurich

1. Introduction

The subject of this paper is to give a survey on algorithms designed for the solution of probabilistic constrained problems with joint probabilistic constraints. A brief summary of the most important convexity results is also included as convexity properties play a central role in developing solution methods for this problem class. For overviews on algorithms aiming the solution of probabilistic constrained problems see also [25], [26] and [50].

The next section contains a brief summary of the probabilistic constrained modeling approach. The third section is devoted to summarize the main results concerning the convexity of the feasible domain of probabilistic constrained problems with joint probabilistic constraints. The subsequent section discusses the numerical difficulties concerning the solution of these problems and considers some implications imposed by this difficulties on algorithm-design. In the fifth section an overview on solution methods is given from the computational point of view. Only algorithms for problems with an absolutely continuous probability distribution are considered and the presentation is restricted to methods which have been implemented on a computer. For the case of a multinormal distribution a new method for finding a feasible solution is also presented in this section. The final section lists some characteristics of the existing implementations of the methods discussed in the previous section.

2. Model formulation

In this section a brief summary will be given on stochastic linear programming models of the probabilistic constrained type. Unfortunately there is no unique terminology for the models discussed below; some authors use the term "probabilistic constrained programming" others "chance constrained programming". To keep connection to the literature both terms will be indicated in the definitions of this section. In subsequent sections the term "probabilistic constrained programming" will be used for models with joint probabilistic constraints. The class of probabilistic (or chance) constrained stochastic programming problems will be introduced directly via the underlying linear programming problem; for a general approach yielding all major stochastic programming model classes see Kall [22]. Let us consider a linear programming problem in the following partitioned form.

$$
\begin{aligned}
&\text{minimize} && c\,x, \\
&\text{subject to} && T^{(k)}\,x \ge h^{(k)}, && k=1,\dots S, \\
& && A\,x = b, \\
& && x \ge 0,
\end{aligned}
\tag{2.1}
$$

---

* on leave from the Computer and Automation Institute, Hungarian Academy of Sciences, Budapest.

where vectors c and x are n-dimensional, $h^{(k)}$ is r_k-dimensional, k=1,...S; b is m-dimensional;

$T^{(k)}$ and A are (r_k x n) and (m x n) matrices, respectively, k=1,...S.

Problem (2.1) becomes a subject of stochastic linear programming in the case when the objective coefficients as well as the entries of the constraint matrices and right-hand-sides of the first partitioned part of the constraints are only available in the statistical sense i.e. they are random variables with a known probability distribution. Let us introduce the following denotation.

$(\Omega, \mathcal{F}, P_\omega)$ is a probability space;

$c(\omega), h^{(k)}(\omega), T^{(k)}(\omega)$, k=1,...S, $\omega \in \Omega$; are random variables and matrices.

A formal replacement of the vectors and matrices through the corresponding random vectors and random matrices in (2.1) results in a meaningless problem formulation. Replacing the random entries in the model by their expected (mean) value leads to a well-defined problem but may produce extremely unreliable solutions, see Prékopa [56] and Kall [22]. In a modeling situation as described above one of the possible ways for getting a well-defined and meaningful problem formulation leads to probabilistic (or chance) constrained problems. Consider the following model.

$$
\begin{array}{lll}
\textit{minimize} & \mathbf{E}_\omega c(\omega) \, x, & \\
\textit{subject to} & \mathbf{P}_\omega(\{\, \omega \mid T^{(k)}(\omega) \, x \geq h^{(k)}(\omega)\,\}) \geq \alpha_k, & \text{k=1,...,S,} \\
& Ax \qquad\qquad\qquad\qquad\quad = b, & \text{(2.2)} \\
& x \qquad\qquad\qquad\qquad\quad\ \geq 0, &
\end{array}
$$

where

α_k, k=1,...,S are prescribed probability (reliability) levels;

it is assumed that the expected value $\mathbf{E}_\omega c(\omega)$ exists.

The modeling approach followed in the probabilistic-constrained framework means that the fulfillment of the random inequalities is prescribed on given probability levels and the expected value of the random objective function is minimized under the resulting constraints. From formulation (2.2) it is clear why the original LP has been considered in a partitioned form: Formulation (2.2) allows for choosing different probability levels for the various blocks by taking into account the appropriate marginal distributions. This is an adequate modeling approach e.g. if the random vectors consisting of the matrix-entries and right-hand-sides of the individual blocks are stochastically independent. For the ease of reference the constraints prescribing probability levels will be referred as probabilistic constraints but notice that (2.2) is a completely deterministic optimization problem. For the sake of simplicity of presentation let us denote the expected value of the vector of objective coefficients again by c and introduce the following denotation for the polyhedron determined by the deterministic linear constraints.

$$\mathcal{P} = \{\, x \mid Ax=b, \, x \geq 0 \,\}$$

This leads to a problem formulation as shown below.

$$
\begin{array}{lll}
\textit{minimize} & c \, x, & \\
\textit{subject to} & \mathbf{P}_\omega(\{\, \omega \mid T^{(k)}(\omega) \, x \geq h^{(k)}(\omega)\,\}) \geq \alpha_k, & \text{k=1,...,S,} \\
& x \qquad\qquad\qquad\qquad\quad \in \mathcal{P}. & \text{(2.3)}
\end{array}
$$

Problem (2.3) is called a <u>probabilistic (or chance) constrained stochastic linear programming</u> <u>problem</u>. In the special case when all of the matrices in the stochastic constraints consist of a single row the model can be formulated as follows.

$$\begin{aligned} \text{minimize} \quad & c\,x, \\ \text{subject to} \quad & \mathbf{P}_\omega(\{\ \omega\mid d^{(k)}(\omega)\,x \geq h_k(\omega)\}) \geq \alpha_k, \qquad k=1,...,S, \\ & x \in \mathcal{P}. \end{aligned} \qquad (2.4)$$

where

$d^{(k)}(\omega)$ denotes the single row of $T^{(k)}(\omega)$, $k=1,...,S$;

$h_k(\omega)$ is the corresponding right-hand-side element, $k=1,...,S$.

In this way <u>probabilistic (or chance) constrained problems with separate probabilistic</u> <u>constraints</u> arise. This way of modeling preserves the option of choosing different probability levels for different rows according to their importance from the reliability point of view. Another special case can be obtained by requiring $S=1$ i.e. allowing only a single probabilistic constraint which results in the following model.

$$\begin{aligned} \text{minimize} \quad & c\,x, \\ \text{subject to} \quad & \mathbf{P}_\omega(\{\ \omega\mid T(\omega)x \geq h(\omega)\}) \geq \alpha, \\ & x \in \mathcal{P}. \end{aligned} \qquad (2.5)$$

where

$T^{(1)}(\omega)$, $h^{(1)}(\omega)$, α_1 were denoted by $T(\omega)$, $h(\omega)$, α, respectively.

Problems of this form are called <u>probabilistic (or chance) constrained problems with joint</u> <u>probabilistic constraints</u>. This approach can be followed in modeling situations where a reliability level should be maintained for the joint fulfillment of the stochastic inequalities. Such models allow also for stochastic dependencies between random variables appearing in different rows. To highlight the difference between the modeling attitudes followed in the separate and in the joint probabilistic constrained approaches let us consider model (2.5) in the case when the random vectors corresponding to different rows are stochastically independent. The probability can be written as a product, and a model different from (2.4) results.

$$\begin{aligned} \text{minimize} \quad & c\,x, \\ \text{subject to} \quad & \prod_{i=1}^{r} \mathbf{P}_\omega(\{\ \omega\mid t^{(i)}(\omega)\,x \geq h_i(\omega)\}) \geq \alpha, \\ & x \in \mathcal{P}. \end{aligned} \qquad (2.6)$$

where

$t^{(i)}(\omega)$ is the i-th row of $T(\omega)$ and $h_i(\omega)$ the i-th component of $h(\omega)$, $i=1,...,r$,

r is the number of rows of matrix $T(\omega)$.

The numerical techniques described in this paper are designed to solve probabilistic (or chance) constrained problems with joint probabilistic constraints, i.e. problems of type (2.5).

Probabilistic (or chance) constrained problems with separate probabilistic constraints were first formulated and investigated by Charnes, Cooper and Symonds [9], [10]. Models with joint probabilistic constraints were introduced by Miller and Wagner [37] for the independent case

(2.6), whereas the general model (2.5) was first formulated and studied by Prékopa [42], [43], [46].

For getting a first idea on the nature of the general problem (2.3) let us consider the case when the technology matrices are deterministic. In this case the model can be written as follows.

$$
\begin{aligned}
\textit{minimize} \qquad & c\,x, \\
\textit{subject to} \qquad & F_k(T^{(k)}\,x) \quad \geq \alpha_k, \qquad k=1,...,S, \\
& x \qquad \in \mathcal{P},
\end{aligned}
\qquad (2.7)
$$

where

$T^{(k)}(\omega)\equiv T^{(k)}$, $k=1,...,S$ are the deterministic constraint matrices,

$F_k(y)$ is the probability distribution function of $h^{(k)}(\omega)$, $k=1,...,S$.

This formulation shows that with the exception of some special cases (see e.g. [22]) probabilistic constrained problems belong to the class of nonlinearly constrained optimization problems. The numerical difficulties with nonlinear optimization problems of the type (2.7) have their roots in the fact that apart from a few special cases the distribution function is not available as a closed algebraic expression but only e.g. as a sum of distribution values in the discrete case or as an integral of the density function in the absolute continuous case. The computation of the probabilistic constraint is usually even more difficult in the general case (2.3). In the framework of stochastic programming (2.3) is considered to be a stochastic linear programming problem, because the deterministic parts of the model are all linear functions of the decision vector x, which also enters the stochastic constraints in a linear way.

For designing efficient solution algorithms for probabilistic (or chance) constrained problems a key issue is the <u>convexity of the feasible domain</u>.

Kall [21] found that the feasible domain is convex, if $\alpha_k=1$, $k=1,...,S$ holds. This fact together with the trivial case $\alpha_k=0$, $k=1,...,S$ are the only available <u>general convexity statements.</u> The points of the feasible set under the assumption $\alpha_k=1$, $k=1,...,S$ were called fat solutions by Madansky [32].

In the case of <u>separate probabilistic constraints</u> the feasible set is obviously convex if randomness appears only on the right-hand-side. This fact is an easy consequence of the monotonicity property of the distribution function, see e.g. [22]. In the general case for separate probabilistic constraints convexity results have been achieved under various assumptions concerning the probability distribution or the problem setting. See Marti [34] and Kall [22] for finite discrete distributions and van de Panne and Popp [41], Marti [34] and Sengupta [63] for continuous distributions.

Convexity of the feasible domain in the case of <u>joint probabilistic constraints</u> is the main subject of the next section.

3. Theoretical basis

This section is devoted to reviewing the main results concerning convexity of the feasible domain of stochastic linear programming problems with joint probabilistic constraints i.e.

problems of the type (2.5). The convexity of the feasible set is clearly implied by the convexity of the following set.

$$X(\alpha) = \{ \ x \ | \ \mathbf{P}_\omega(\{ \ \omega \ | \ T(\omega) \ \ x \geq h(\omega)\}) \ \geq \alpha \ \}. \tag{3.1}$$

In fact problem (2.5) may be reformulated as shown below from which this assertion easily follows.

$$\begin{aligned} &minimize \quad c \ x \\ &subject \ to \quad x \in \mathcal{P} \cap X(\alpha). \end{aligned} \tag{3.2}$$

The discrete case will be considered first. Let us assume that the joint probability distribution is a <u>finite discrete distribution</u> given as follows.

$T(\omega_i)$, $h(\omega_i)$, $i=1,...,K$ are the realizations with positive probabilities,

p_i, $i=1,...,K$ are the corresponding probabilities fulfilling the relations

$$p_i > 0, \ i=1,...,K, \qquad \sum_{i=1}^{K} p_i = 1.$$

Marti [34] found the following sufficient condition for the convexity of the feasible domain.

Let i_0 be an index for which $p_{i_0} = min \ \{ \ p_i \ | \ i=1,...,K \ \}$ holds,

then $\alpha > 1-p_{i_0}$ implies the convexity of $X(\alpha)$.

The result follows from the fact that $\alpha > 1-p_{i_0}$ is equivalent to prescribing $\alpha=1$ in (3.1). The condition cannot be weakened (see [22]); a sharper result was given by Kall [22] for the case when the minimal index above is uniquely determined. Let us assume that this condition is fulfilled then the following proposition holds.

Let i_1 be an index for which $p_{i_1} = min \ \{ \ p_i \ | \ i=1,...,K; \ i \neq i_0 \ \}$ is fulfilled,

in this case $\alpha > 1-p_{i_1}$ implies the convexity of $X(\alpha)$.

The conditions implying convexity impose rather strong requirements on the reliability level as a modeling parameter if the number of realizations is large. If e.g. $K=100$ then the probability level must be greater than 0.99 to ensure convexity. The feasible domain is in general nonconvex which can easily be seen by considering the structure of the set of feasible solutions. Let us introduce the following denotation for the convex polyhedra corresponding to the individual realizations.

$$K(\omega_i) = \{ \ x \ | \ T(\omega_i)x \geq h(\omega_i) \ \} \quad i=1,...,K.$$

The following relation clearly holds.

$$X(\alpha) = \bigcup_{I \in \mathcal{n}} \ \bigcap_{i \in I} K(\omega_i) \tag{3.3}$$

where $\mathcal{n} = \{ \ I \ | \ I \subset \{1,...,K\}, \ \sum_{i \in I} p_i \geq \alpha \ \}$.

Relation (3.3) shows that $X(\alpha)$ can be a represented as a union of convex polyhedra, which means that problem (2.5) contains so-called disjunctive constraints, see [39], implying that the

146

feasible domain is in general not a convex set. This nature of the problem admits reformulations as mixed-variable binary optimization problems. Such a reformulation of (2.5) was first given by Raike [59] having following form.

$$\text{minimize} \quad c\,x,$$

$$\text{subject to} \quad T(\omega_i)\,x + M(1 - z_i)\mathbf{1} \geq h(\omega_i), \quad i=1,...,K,$$

$$\sum_{i=1}^{K} p_i\, z_i \geq \alpha, \tag{3.4}$$

$$x \in \mathcal{P},$$

$$z_i = 0 \text{ or } 1,$$

where
1 is a vector with all of its components equal to 1;
M is to be chosen big enough, see [59] .

Prékopa [50] gave a reformulation of (2.5) without a "big-M" parameter for the case of a deterministic technology matrix T, as follows.

$$\text{minimize} \quad c\,x,$$

$$\text{subject to} \quad T\,x - \sum_{i=1}^{K} h(\omega_i)\, z_i \geq 0$$

$$\sum_{i=1}^{K} F(h(\omega_i))\, z_i \geq \alpha, \tag{3.5}$$

$$\sum_{i=1}^{K} z_i = 1,$$

$$x \in \mathcal{P},$$

$$z_i = 0 \text{ or } 1,$$

where $F(y)$ is the probability distribution function of the random right-hand-side. Both problems (3.4) and (3.5) are linear mixed-variable optimization problems with as many Boolean variables as many joint realizations the random variables of the problem have.

As a second class absolutely continuous probability distributions will be considered, i.e. distributions for which the probability measure is generated through a density function. If a random technology matrix is allowed in (2.5) then convexity of the feasible domain can only be assured under quite strong assumptions even in the case of a multinormal distribution, see Prékopa [47] and Burkauskas [7], [8]. The latter author also found some convexity results for the broader class of multivariate stable distributions, see [7] and [8]. In the sequel it will be assumed that the technology matrix is deterministic, i.e. that

$$T(\omega) \equiv T$$

holds. In this case (2.5) can be reformulated as given below.

$$\text{minimize} \quad c\,x,$$

$$\text{subject to} \quad F(T\,x) \geq \alpha,$$

$$x \in \mathcal{P}, \tag{3.6}$$

where F(y) is the probability distribution function of the random right-hand-side h(ω). The reformulation of (2.5) can be carried one step further by introducing new variables (see Miller and Wagner [37]) resulting in the problem as follows.

$$
\begin{aligned}
\text{minimize} \quad & c\,x, \\
\text{subject to} \quad & F(y) \geq \alpha, \\
& Tx - y \geq 0, \\
& x \in P.
\end{aligned} \tag{3.7}
$$

The equivalence of (3.6) and (3.7) is an easy consequence of the monotonicity of the distribution function. From (3.7) it is evident that the quasiconcavity of the distribution function implies the convexity of the feasible domain. For giving an overview of results concerning the quasiconcavity of distribution functions the following notion will be needed.

Definition.
    A real-valued function f defined on the whole space is said to be α-concave, if for any two points x,y and any $0 \leq \lambda \leq 1$ the following inequality holds.

$$
f(\lambda x+(1-\lambda)y) \geq m_{\lambda,\alpha}(f(x),f(y)) \tag{3.8}
$$

    A probability measure P is said to be α-concave, if for any Borel-measurable convex sets A,B,C with $C=\lambda A+(1-\lambda)B$ and $0 \leq \lambda \leq 1$ the inequality $P(C) \geq m_{\lambda,\alpha}(P(A),P(B))$ holds.

    In the definitions above $m_{\lambda,\alpha}(p,q)$ is the well-known mean introduced by Hardy, Littlewood and Polya see e.g. [4]. From a monotonicity property of this mean it readily follows that if $\alpha_1 < \alpha_2$ holds then α_2-concavity implies α_1-concavity.

    From the stochastic programming point of view the most important special case corresponds to α=0 defining the class of logarithmic concave (also called logconcave) functions and measures. In this case (3.8) takes the following form.

$$
f(\lambda x+(1-\lambda)y) \geq f(x)^\lambda f(y)^{(1-\lambda)} \tag{3.9}
$$

For the interpretation of this type of generalized concavity in terms of the usual concavity notion see Figure 1.

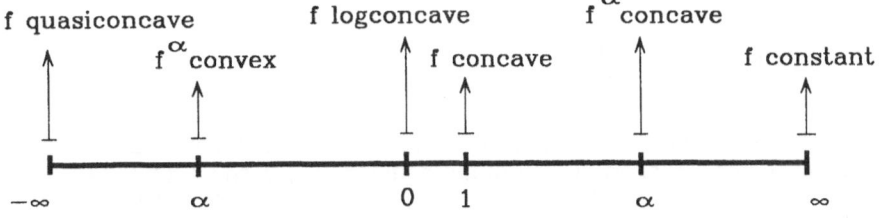

Figure 1. α - concavity of a function f.

For the α-concavity of the distribution function it is sufficient to prove the α-concavity of the probability measure. This fact can be proved e.g. using the same argument as Prékopa used in [49] for the logarithmic concave case.

A generalized concavity result concerning measures was first obtained by Brunn, Minkowski and Lusternik, see [3] who proved that the Lebesgue-measure is 1/n-concave. Prékopa [43], [45] recognized the importance of generalized concave measures in the stochastic programming framework and proved that distribution functions belonging to a broad class are logarithmically concave. His theory of logarithmic concave measures resulted in a breakthrough in the field of probabilistic constrained programming. Borell [4] extended the theory in the sense of inequality (3.8) and proved quasiconcavity of a larger class of distribution functions (see also Figure 2). The main result concerning generalized concave measures can be formulated as follows.

If the probability density function is
α-concave for $-1/n \leq \alpha \leq +\infty$,
then the probability measure and consequently the distribution function is

$$\frac{\alpha}{1+n\alpha} \text{ - concave.}$$

For $\alpha=+\infty$ this was proved by Brunn, Minkowski and Lusternik, see [3];
for $\alpha=0$ by Prékopa [43]; see also [42]; [46]; [49];
extended to the whole range by Borell [4].

For different proofs and extensions see also [61], [6]. Inverse results (generalized concavity of the measure implies generalized concavity of the density function) have been found by Borell [4], Brascamp and Lieb [6] for α-concave measures, and by Kall [22] for the special case of quasiconcave measures.

In Figure 2 a schematic representation of the above results can be seen along with the main distribution-classes having a generalized concavity property. For some of the distributions the displayed in Figure 2 generalized concavity holds only on a subset of the possible parameter values, see [4] and [43]; for the particular distributions see [20]. In Figure 2 the t-, F- and Pareto-distribution functions are indicated to be -1/n -concave. This is certainly true but actually they are α -concave with an α depending on the parameters of the distributions. For all possible values of the distribution-parameters the corresponding α belongs to the interval (-1/n, 0) thus implying -1/n -concavity of the density function; see [4].

Remark. Having the logconcavity property for a distribution function means much better behavior from the mathematical programming point of view than just having quasiconcavity. It is easy to see that logconcavity implies strict quasiconcavity and in the differentiable case even pseudoconcavity. For the generalized concavity notions just mentioned see e.g. [1].

4. Numerical considerations

The subject of this section is to consider probabilistic constrained problems from the algorithmic point of view. A short survey of the sources of numerical difficulties will be given along with the requirements they impose on solution algorithms. We will confine ourselves to the case of a deterministic technology matrix in the probabilistic constraint because all available general-purpose algorithms which aim to solve (2.5) are designed for this case. This restriction implies that the model may be considered in the form given in (3.6) and (3.7). The main purpose of this paper being to give an overview on solution methods for (2.5) with an absolutely continuous distribution only this class of problems will be considered.

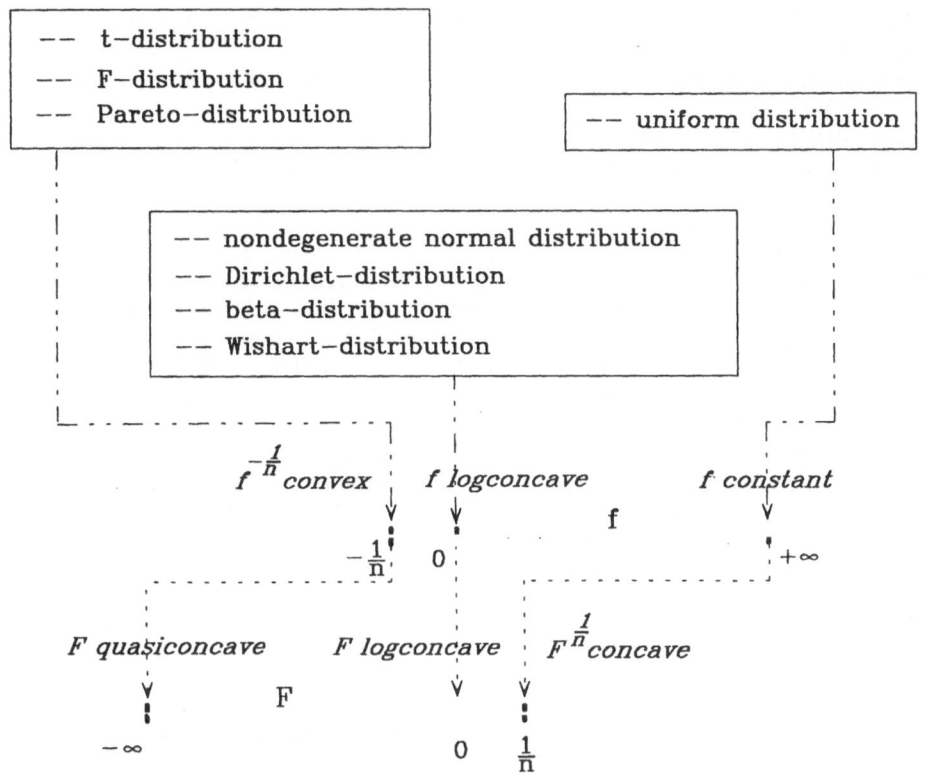

Figure 2. α - concavity of some distribution and density functions.

For probabilistic constrained problems with a finite discrete distribution Prékopa recently gave a dual-type algorithm and reports on very encouraging computational results [54]. He created an APL implementation of his method and a FORTRAN implementation was created by the graduate student A. Burnetas.

In the sequel it will be assumed that the random variables in (2.5) have an absolutely continuous joint probability distribution.

(1) The computation of F(y) means the evaluation of multidimensional integrals which is a time-consuming task on the computer. All implemented solution algorithms rely on Monte-Carlo integration for computing F(y) which seems to be the only feasible approach for problems with dimension >3 in the random right-hand-side. In connection with probabilistic constrained problems the following simulations schemes have been used: the Monte-Carlo technics of Deák [11], [12], [13]; Szántai's scheme [64], [65] which combines simulation with numerical integration and Gassmann's method [18] which is a combination of the schemes of Szántai and Deák. A very promising approach based on quasi-Monte-Carlo methods has been developed by Niederreiter, see e.g. [40].

Standard nonlinear optimization techniques frequently call for values of F(y) in the iterative solution process. Feature (1) implies that they should be modified to avoid the computation of F(y) as much as possible which can be achieved e.g. by utilizing cheaply computable bounds for F(y). Such bounds are available via the Boole-Bonferroni inequalities, see Prékopa [51], [52], [53], Boros and Prékopa [5] and Szántai [65].

(2) If $\nabla F(y)$ exists and an algorithm uses this gradient then the question arises how to compute it. Theoretically under mild assumptions the gradient can be computed using the conditional distribution function. Practically this usually means again Monte-Carlo integration, see Prékopa [49].

In the existing solution procedures for (3.6), reducing the number of computations of the gradient seems to be difficult; using stochastic quasigradient methods (see [15], [17]) might be a possibility to overcome this difficulty.

(3) The computation of $F(y)$ and its gradient can only be performed with a limited accuracy in reasonable computing time via the Monte-Carlo approach. This means that computations involving the probabilistic constraint introduce a numerical error several magnitudes higher when compared with the deterministic parts of the model. Another source of inaccuracy is the modeling error which is usually high due to the fact that the parameters of the distribution are determined with the help of some statistical procedures.

This means that it has no sense to require a solution of (3.6) with high accuracy, an accuracy between 1% and 5% is quite sufficient in most cases. All implemented solution algorithms for (3.6) are first-order methods from the point of view of nonlinear programming; it has not yet been explored whether anything could be gained by implementing also second-order methods as they usually speed up in a close neighborhood of the optimal solution.

(4) $F(y)$ is not a concave function; for most distributions it is either logconcave or just quasiconcave.

Qasiconcavity of $F(y)$ implies the convexity of the feasible domain, but solution algorithms must carefully be designed to ensure convergence under appropriate regularity assumptions. Having a quasiconcave distribution which is not logconcave may cause serious difficulties in the Phase-I procedure if this relies on maximizing $F(y)$.

(5) Model (3.6) is considered as an extension of the linear programming modeling approach.

This feature implies that no special size-limitations should be imposed on the deterministic part of the problem, in this respect large-scale problems should be allowed for. For designing solution algorithms this means that extensive use should be made of the well-developed and readily-available techniques designed for large-scale linear programming. A desirable property of a solution algorithm in this respect is a separate handling of the probabilistic constraint and the LP-part of the problem.

5. Solution techniques

This section is devoted to discussing the main features of algorithms designed for the solution of (2.5). It will be pointed out how the separate algorithms try to overcome the difficulties considered in the previous section. The main emphasis will be laid on the explanation of the basic ideas underlying the algorithms. This means that some important details, e.g. stopping criteria, will not be considered. Concerning details the interested reader is referred to the literature. All of the algorithms considered here rely on some well-known principle or algorithmic framework of nonlinear programming and they will be addressed according to this underlying principle. In all cases however they were modified to utilize specialities of (2.5) resulting in new algorithms or algorithm-variants which may be attributed to the authors. The following algorithms will be reviewed: A feasible direction method, logarithmic barrier methods, a supporting hyperplane method, a dual method and a reduced gradient method. Some of the algorithms require a starting feasible point; the different approaches developed for finding such a starting point will be summarized in a separate subsection. The promising direction of applying approximation schemes will shortly be addressed in the final subsection. The individual algorithms are discussed from the computational point of view but it is not our intention to suggest any conclusions concerning efficiency of the methods or even a ranking between them. According to our opinion these issues may only be addressed on the basis of an extensive comparative computational study.

For the sake of simplicity of presentation let us introduce the following denotation.

$$G(x) = F(Tx).$$

Using this problem (3.6) takes the following form.

$$
\begin{aligned}
minimize \quad & c\,x, \\
subject\ to \quad & G(x) \geq \alpha, \\
& x \in \mathcal{P}.
\end{aligned}
\tag{5.1}
$$

5.1. A feasible direction method

Historically the first algorithm which was successfully applied to the solution of probabilistic constrained problems was the method of Prékopa and Deák [56] based on a feasible direction method of Zoutendijk. The algorithm generates a sequence of feasible points with improving objective values by determining at each iteration first a feasible direction, then a feasible stepsize and taking as the next feasible point the point obtained by performing the step. Let us assume that an initial feasible point is available, and that at the beginning of the N-th iteration a feasible point x^N is given. The next point will be computed as follows.

Step 1. Solve the following direction finding subproblem.

$$
\begin{aligned}
minimize \quad & \zeta, \\
subject\ to \quad & c(x-x^N) - \zeta \leq 0, \\
& \nabla G(x^N)(x-x^N) + \Theta\zeta \geq 0, \quad \text{if } G(x^N)=\alpha, \\
& x \in \mathcal{P}.
\end{aligned}
\tag{5.2}
$$

where the parameter Θ is a fixed positive constant serving as a scaling factor. If the optimal objective value in (5.2) is 0 then the current feasible solution is optimal, otherwise denote the x-part of the optimal solution by x^*.

Step 2. Linesearch. The stepsize is the maximal steplength for which the new point is still feasible when moving from the previous solution into the direction of the optimal solution of (5.2) and can be obtained as follows.

$$\lambda^* = max\{\ \lambda \mid x^N+\lambda\,w \in \mathcal{P},\ G(x^N+\lambda\,w)\geq\alpha\ \};\qquad w=(x^*-x^N). \tag{5.3}$$

Convergence: The convergence of the algorithm has been proved by Prékopa [48] under the following assumptions: Quasiconcavity and continuous differentiability of G, boundedness of the polyhedron determined by the deterministic linear constraints, and a regularity condition for (5.1).

Numerical aspects.

The direction-finding subproblem (5.2) is a linear programming problem and may be solved by using any advanced LP-package. It is advisable to utilize information concerning the optimal solution of the previous iteration to provide an advanced start for the current iteration.

The computation of the steplength as formulated in (5.3) involves the determination of the intersection of a ray with the boundary determined by the probabilistic constraint. The implementation of this operation plays a crucial role in the overall numerical performance of almost all of the algorithms discussed in this paper. For this feasible-direction algorithm an iterative procedure based on forward-, and backward-movements along the ray governed by computed values of F(y) has been proposed in [56]. This subprocedure determines the intersection within a fixed tolerance level and achieves additionally that the intersection point is feasible with respect to the probabilistic constraint. The computing time is reduced in the following way. G(x) is first computed with a low accuracy by selecting a relatively small sample-size in the Monte-Carlo procedure. If using the predicted error a safe decision can be made whether the current point is feasible with respect to the probabilistic constraint then the subprocedure performs its next step, otherwise G(x) will be computed more accurately by increasing the sample size.

5.2. Logarithmic barrier (SUMT) methods

For the case of a logconcave distribution function Prékopa [44] proposed to use logarithmic barrier functions in the SUMT [16] computational scheme. This approach results in a concave objective function in the problem for which unconstrained minimization techniques are applied. The idea has been implemented by Rapcsák [60] and Prékopa and Kelle [57] to solve probabilistic constrained real-life problems. No general-purpose code has been developed. The two methods differ by using different unconstrained minimization techniques, which were in both cases direct-search methods.

The main idea in using logarithmic barrier methods to solve (3.6) is solving a sequence of problems of the following form.

$$\begin{array}{ll} minimize & cx-R \log(G(x)-\alpha) \\ subject\ to & x \in P. \end{array} \qquad (5.4)$$

where the penalty parameter R>0 is decreased according to specific strategies (see e.g. [1], [16] . Problem (5.4) is solved by algorithms for linearly constrained nonlinear programming problems, where special attention must be given to the steplength-determination to avoid "jumping over" the barrier by using too large steps. The objective function in (5.4) is in fact concave because the translation preserves logconcavity, see [50].
Convergence. Problem (5.4) being a convex programming problem the usual convergence results may be applied, see [1].
Numerical aspects. The numerical behavior highly depends on the algorithm implemented for the solution of the linearly constrained nonlinear problem (5.4) where additionally special care must be taken concerning stepsize control.

5.3. A supporting hyperplane method

This algorithm, based on Veinott's supporting hyperplane method (see e.g. [1]) was developed by Szántai [66] and used by Prékopa and Szántai [58] to solve a reservoir design problem of Prékopa. Two sequences of points are generated; one of them consists of feasible points satisfying strictly the probabilistic constraint (Slater-points), the other consists of infeasible points. Let us assume that at start the following quantities are given.

$z^1 \in P$, for which $G(z^1) > \alpha$ holds,

$y^1 \in P$, for which $G(y^1) < \alpha$ holds, (Slater-point), and

let $L^1 = P$.

Assume that z^N, y^N with the properties above are given along with L^N, then the next iteration consists of the following steps.

Step 1.Solve the following linear programming problem.

$$\begin{aligned} \text{minimize} \quad & cx, \\ \text{subject to} \quad & x \in \mathcal{P} \cap \mathit{l}^N \end{aligned} \tag{5.5}$$

and denote the optimal solution of (5.5) by z^{N+1}. If $z^{N+1} \in \mathcal{P}$ then it is an optimal solution of (5.1), otherwise continue with Step 2.

Step 2. Let x^{N+1} be the intersection of the line-segment joining z^{N+1} and y^N with the surface

$$\{ x \mid G(x) = \alpha \}. \tag{5.6}$$

Step 3. Perform a cut using the supporting hyperplane based on a linearization of $G(x)$ to get a new polyhedron in (5.5).

$$\mathit{l}^{N+1} = \mathit{l}^N \cap \{ x \mid \nabla G(x^{N+1})(x-x^{N+1}) \geq 0 \}. \tag{5.7}$$

It is easy to see that the optimal solution of (5.5) has been cut off. i.e. $z^{N+1} \notin \mathit{l}^{N+1}$ holds.

Step 4. Modify the Slater-point by moving it closer to the boundary.

$$y^{N+1} = y^N + \frac{1}{N}(x^{N+1}-y^N). \tag{5.8}$$

In the original Veinott-method the Slater-point is kept fixed; in a variant proposed by Zoutendijk (5.8) is replaced by

$$y^{N+1} = y^N + \beta(x^{N+1}-y^N), \tag{5.9}$$

with a fixed proportion β, $0<\beta<1$. Szántai found by computational experimentation (5.8) to be the best strategy [66]. Let us turn attention to some similarities between this method and the feasible direction method discussed in subsection 5.1. The second constraint in the direction-finding subproblem (5.2) just cuts off the previous feasible solution generated by the method through the same kind of linearization of the probabilistic constraint as it is used in (5.7). In fact the supporting hyperplane method can be considered as a modification of Zoutendijk's feasible direction method described in 5.1. The modification involves accumulation of the cutting constraints and accelerating the algorithm by utilizing a moving Slater-point; see Zoutendijk [67].

Convergence. A convergence proof exists only for the original version of Veinott's method with a fixed Slater-point under the assumptions of the pseudoconcavity of F, boundedness of the set determined by the linear constraints and Slater-condition for (5.1). For the version of the supporting hyperplane method described above no convergence proof is known.

Numerical aspects.

Step1 involves the solution of a linear programming problem, any one of the existing powerful LP-packages can be used. As with the feasible direction method it is advisable to use information from the previous iteration to have a good start in the current iteration.

In Step 2 of the algorithm the same task is to be performed as in Step 2 of the feasible direction method namely to compute the intersection of a line with the surface (5.6) determined by the probabilistic constraint. Szántai uses a bisection approach which relies on cheaply computable bounds based on Boole-Bonferroni inequalities, see [65], for the case of the multinormal distribution. The intersection is located within a prescribed tolerance which is kept fixed during the procedure, and a slight infeasibility is enforced to avoid cutting off parts of the feasible domain.

5.4. A dual method

The probabilistic constrained problem will be considered in the form (3.7) which can be formulated with explicitly given deterministic linear constraints as follows.

$$
\begin{array}{ll}
\text{minimize} & c\,x, \\
\text{subject to} & F(y) \geq \alpha, \\
& Tx - y \geq 0, \\
& Ax = b, \\
& x \geq 0.
\end{array}
\tag{5.10}
$$

A Lagrange-dual to (5.10) in the sense of Geoffrion [19] can be formulated as shown below.

$$
\begin{array}{ll}
\text{maximize} & \psi(u)u + b\,v, \\
\text{subject to} & T'u + A'v \geq c, \\
& u \geq 0,
\end{array}
\tag{5.11}
$$

where $\psi(u)$ is the optimal solution of the following optimization problem,

$$
\begin{array}{ll}
\text{minimize} & u\,y, \\
\text{subject to} & F(y) \geq \alpha,.
\end{array}
\tag{5.12}
$$

and T', A' denote the transposed matrices.

This form of the dual of probabilistic constrained problems has been given by Komáromi [30]; her dual method is based on the duality relationship between (5.10) and (5.11). Let us remark that Luc [31] constructed a dual via conjugate-functions for the general probabilistic constrained problem where the technology matrix in the probabilistic constraint may also be random. His dual specializes to (5.11) in the case of a deterministic technology matrix. The dual algorithm of Komáromi [30] works as follows. Assume that a feasible solution (u^N, v^N), $N \geq 1$ of the dual problem (5.11) is available, then the next iteration consists of the following steps.

Step 1. Determine $\psi(u^N)$ by solving (5.12) with $u = u^N$.

Step 2. Solve the linear programming problem given below.

$$
\begin{array}{ll}
\text{maximize} & \psi(u^N)u + b\,v, \\
\text{subject to} & T'u + A'v \geq c, \\
& u \geq 0,
\end{array}
\tag{5.13}
$$

let the solution be denoted by $(\tilde{u}^N, \tilde{v}^N)$

Case 1.

$$
\exists \tau: \tilde{u}^N = \tau u^N
\tag{5.14}
$$

holds. In this case $\psi(u^N)$ is the y-part of an optimal solution of (5.10). By fixing $y = \psi(u^N)$ in (5.10) a linear programming problem results for the x-part of the optimal solution.

Case 2. Relation (5.12) does not hold. In this case proceed with the next step.

Step 3. Linesearch. The vector pointing from (u^N, v^N) into the direction of $(\tilde{u}^N, \tilde{v}^N)$ is a direction of ascent for the dual objective. A stepsize λ_N is determined, see [30], and the new dual feasible solution computed as

$$
(u^{N+1}, v^{N+1}) = (u^N, v^N) + \lambda_N (\tilde{u}^N - u^N, \tilde{v}^N - v^N)
\tag{5.15}
$$

<u>Convergence</u>. Convergence of the dual method has been proved by Komáromi [30] under the following assumptions: Strict logarithmic concavity and continuous differentiability of F, a regularity condition and the application of a certain stepsize-rule in Step 3 which involves the function $\psi(u)$, see [30].

<u>Numerical aspects</u>.

The nonlinearly constrained optimization problem (5.12) is to be solved at each iteration in Step 1. The only difference between this problem and the original problem (5.10) is that linear constraints are missing in (5.12). An efficient method should be developed for the solution of (5.12) which utilizes the absence of the linear constraints, [30] does not contain any clue in this respect.

Step 2 of the method can efficiently be implemented because only the objective of the linear programming problem changes between two iterations.

To utilize theoretical convergence the stepsize should be determined according to a rule given in [30] which contains the function $\psi(u)$. Practically this means repeated solution of (5.12) again.

5.5. A reduced gradient method

The problem will be considered in the following form.

$$
\begin{aligned}
\text{minimize} \quad & c\,x, \\
\text{subject to} \quad & G(\,x\,) \geq \alpha, \\
& A\,x = b, \\
& x \geq 0,
\end{aligned}
\tag{5.16}
$$

where $G(x) = F(Tx)$.

The algorithm belongs to the class of feasible direction methods where the deterministic linear constraints are handled by a reduced gradient technique. The method was developed by Mayer [35], [36]. Let us assume that at the beginning of the N-th iteration, $(N \geq 1)$, we are given a feasible solution x^N of (5.14), a tolerance $\varepsilon^N > 0$, and a partitioning $x=(y^N,z^N)$, $A=(B,N)$, $c=(g,h)$ fulfilling the following relations.

B is a nonsingular square matrix;

$(y^N)_j > \varepsilon^N$, $(\forall j)$ holds (nondegeneracy).

The next iteration consists of the following steps.

<u>Step 1</u>. Solve the direction-finding subproblem to determine a direction $w=(u,v)$.

Consider a direction finding subproblem for (5.16) in the following form.

$$
\begin{aligned}
\text{minimize} \quad & \zeta, \\
\text{subject to} \quad & g\,u + h\,v \leq \zeta, \\
& \nabla_y G(x^N)u + \nabla_z G(x^N)v \geq \Theta\zeta, \quad \text{if } G(x^N) \leq \alpha + \varepsilon^N, \\
& B\,u + N\,v = 0, \\
& v_j \geq 0, \quad \text{if } z_j \leq \varepsilon^N \\
& \|v\|_\infty \leq 1.
\end{aligned}
\tag{5.17}
$$

with Θ being a positive weighting factor.

Under our assumptions problem (5.17) can be reduced to the problem given below.

$$
\begin{aligned}
\text{minimize} \quad & \zeta, \\
\text{subject to} \quad & r\,v \leq \zeta, \\
& s\,v \geq \Theta\zeta, \quad \text{if } G(x^N) \leq \alpha + \varepsilon^N, \\
& v_j \geq 0, \quad \text{If } z_j \leq \varepsilon^N \\
& \|v\|_\infty \leq 1.
\end{aligned}
\tag{5.18}
$$

where

$r' = h' - g'B^{-1}N$, and $s' = \nabla_z G(x^N) - \nabla_y G(x^N)B^{-1}N$ are the reduced gradients of the objective and the probabilistic constraint, respectively. Let the optimal solution of (5.18) be denoted by (v^*, ζ^*).

Case 1. $\zeta^* > \varepsilon^N$; the u-part of the direction w can be computed using the relation

$$u^* = -B^{-1}Nv^*, \tag{5.19}$$

and Step 2. follows.

Case 2. $\zeta^* \leq \varepsilon^N$; in this case ε^N is halved. If the resulting ε^N is smaller than a zero tolerance then x^N is accepted as an optimal solution, otherwise (5.18) is solved again now with the new ε^N.

Step 2. Linesearch. To get the steplength determine the intersection of the ray determined by w and the boundary of the feasible domain.

$$\lambda_1 = max \{ \lambda \mid x^N + \lambda w \in \mathcal{P} \},$$
$$\lambda_2 : \quad \alpha \leq G(x^N + \lambda_2 w) \leq \alpha + \varepsilon^N, \tag{5.20}$$
$$\lambda^* = min \{ \lambda_1, \lambda_1 \}.$$

Compute the new feasible point: $x^{N+1} = x^N + \lambda^* w$.

Step 3. Change the basis. If $(\exists j)$: $(y^{N+1})_j \leq \varepsilon^N$ then the nondegeneracy assumption is violated. Find a new partition for which the nondegeneracy assumption holds; reduce ε^N if this is necessary to fulfill the nondegeneracy requirement.

Remark. (Degeneracy) In Step 3 it may happen that there exists no basis with the required property. In this degenerate case it is always possible to find a basis having the following property: For the direction computed by (5.18), (5.19) $y_i=0$ implies $u_i \geq 0$: The algorithm proceeds by using this direction. The idea is to work with a maximal basis (see Kleinmichel and Sadowsi [29] and Dembo and Klincewicz [14] for the linearly constrained case). A basis is called maximal if there exists no other basis with more strictly positive variables. Such a basis can always be found using a greedy strategy. Having a maximal basis the required basis is found by performing dual pivot steps on the original direction finding problem (5.17) where nonnegativity for the direction-components is prescribed also for degenerate basic components and the norm-constraint involves the whole direction vector w.

Convergence. The convergence has been proved in [35] under the following assumptions: G(x) is logconcave, $\nabla G(x)$ is Lipschitz-continuous, the feasible domain is bounded and the Slater-condition holds.

Numerical aspects.

To perform an iteration of the algorithm it is sufficient to keep and update B^{-1}, the algorithm can be built up on the basis of a modular linear programming system e.g. XMP (see Marsten [33])

The direction-finding subproblem (5.18) is a linear continuous knapsack problem, its solution involves essentially just sorting.

In Step 2 the surface determined by the probabilistic constraint is to be located within a tolerance limit which is gradually decreased during the procedure. This gives good chances for avoiding the computation of G(x) by using bounds. In fact a bisection strategy is used where for the case of the multinormal distribution Szántai's procedure [65] for computing bounds is utilized.

Building-in second-order methods can be performed in the same way as by MINOS (See Murtagh and Saunders [38]). Nonbasic variables are subdivided into free and fixed variables, the direction-finding subproblem (5.18) contains only free variables and the reduction is to be carried out one step further in (5.18).

5.6. Finding a starting feasible solution

A Slater-point can be found by applying a nonlinear programming algorithm designed to solve linearly constrained nonlinear optimization problems to the problem given below, see Prékopa [56].

$$\begin{array}{ll} maximize & G(x), \\ subject\ to & x \in \mathcal{P}. \end{array} \qquad (5.21)$$

This general approach is always applicable if $F(y)$ is logarithmically concave. If a Slater-point exists then finding such a point can be performed in a finite number of steps by any convergent algorithm. As a starting point the optimal solution of the underlying linear programming problem can e.g. be used in which the random right-hand-side is replaced by expected values as shown below.

$$\begin{array}{ll} minimize & c\,x, \\ subject\ to & T\,x \geq \bar{h}, \\ & x \in \mathcal{P}, \end{array} \qquad (5.22)$$

where $\bar{h}=\mathbf{E}_\omega h(\omega)$.

For the case of the multinormal distribution Szántai [66] proposed the following heuristics. He replaces $\bar{h}$ in (5.22) by the right-hand-side given below.

$$\bar{h}=\mathbf{E}_\omega h(\omega)+t\sigma_\omega h(\omega),\ \text{with } \mathbf{E} \text{ denoting expected value and } \sigma \text{ standard deviation.}$$

The objective is also changed, see [66]. The heuristics for finding a starting feasible point is based on solving (5.22) repeatedly with the right-hand-side changed by changing the parameter t. As a starting value $t=3$ is recommended.

Szántai's approach can be improved by taking into account also the covariances. It is well-known that the following random variable is chi-square distributed with r degrees of freedom (with r being the dimension of the random right-hand-side).

$$U=(h(\omega)-\mu)\Sigma^{-1}(h(\omega)-\mu), \qquad (5.23)$$

where $\mu=\mathbf{E}_\omega h(\omega)$, and Σ denotes the covariance-matrix.

Let us consider the ellipsoid given below.

$$E=\{\,y \mid (y-\mu)\Sigma^{-1}(y-\mu)\leq d^2\,\}. \qquad (5.24)$$

Using the chi-square distribution it is possible to chose for any α the constant d in such a way that the probability-content of E will be 1-α. Having determined E the solution of the following optimization problems determine a right-hand-side $\bar{h}$.

$$\bar{h}_i: \qquad \begin{array}{ll} maximize & y_i, \\ subject\ to & y \in E. \end{array}$$

These optimization problems are explicitly solvable e.g. by using Cholesky-factorization. Geometrically the negative orthant is shifted till it contains the whole ellipsoid. Notice that the points computed this way lay on a straight line passing through the center of the ellipsoid for different values of d; this fact allows for a parametrization like that in Szántai's method.

5.7. Approximation schemes

The main numerical difficulties concerning probabilistic constrained problems have all their roots in the need to evaluate multidimensional integrals.

One way to overcome this difficulty is to give easily computable approximations to the corresponding probabilities. Based on Boole-Bonferroni inequalities Prékopa [51], [52] gave a computationally attractive linear programming approach to compute approximations to the probabilities which appear in probabilistic constraints. The approach has been extended and further elaborated by Prékopa [53], Boros and Prékopa [5]. For the special case of a stochastic transportation network see Prékopa and Boros [55].

Another possibility could be using approximations to the measure which result in easily computable integrals. In fact approximating the distribution through finite discrete distributions is an efficient approach for solving two-stage problems and may turn out to be a computationally feasible approach for probabilistic constrained problems as well. For general approximation-schemes of this kind see [2], [23], [24], [26], [27]. Especially for probabilistic constrained problems Salinetti [62] designed an approximation scheme for solving probabilistic constrained problems through solving a sequence of problems with finite discrete distributions. The success of the discrete-approximation approach largely depends on the development of methods for solving problems with finite discrete distributions. The availability of a recent method of Prékopa [54] for this problem class will certainly stimulate the development of algorithms of the discrete approximation type for probabilistic constrained problems.

6. Implementations

No systematic comparative study has been performed concerning the computational efficiency of the algorithms discussed in the previous section. The only test problem which has been solved by the majority of the algorithms is the STABIL model, see Prékopa et al [56] and King [28]. Szántai published a further test problem in [66]. The individual methods have been implemented on different computers, they are utilizing different kinds of Phase-I procedures, use different LP-systems, employ different procedures to compute F(y) and its gradient, and solve the problem with different accuracy. Under such circumstances we confine ourselves to just listing with the methods some of these characteristic features of the implementation. For more details see the references in the previous section. Giving some idea of performance will only be done in connection with our own method.

(1) <u>Feasible direction method</u>.
> Computer used: CDC 3300 mainframe.
> Phase-I: Maximizing the probability using a feasible-direction method.
> Computing F and ∇F: Monte-Carlo integration, see Deák [13].
> Problem solved: STABIL [56]. .

(2) <u>Logarithmic barrier methods</u>.
> Computer used: CDC 3300 mainframe.
> Phase-I: No starting feasible solution needed.
> Computing F and ∇F: Monte-Carlo integration, see [60], [57].
> Problems solved: Reservoir-design [60] and inventory control [57] problems.

(3) <u>Supporting hyperplane method</u>.
> Computer used: UNIVAC 1108 mainframe and IBM/PC AT.
> Phase-I: Heuristic methods as discussed in the previous section.
> Computing F and ∇F: Monte-Carlo integration, see [65].
> Problems solved: STABIL [56], reservoir-design problems [58] and a problem

described in [66].

(4) <u>Dual method</u>.
> Computer used: IBM 3031 mainframe.
> Phase-I: A starting dual feasible solution is computed by using MPSX.
> Computing F and ∇F: The method has only been implemented for random right-hand-sides with independent components and for 2-dimensional random right-hand-sides. The first case is trivial from the point of view of computing F and its gradient, in the second case probably numerical quadrature has been used (it is not clear from the paper), anyway the author reports on starting the procedure by tabulating values of the 2-dimensional normal distribution and using this tabulated values afterwards, see [30].
> Problems solved: Test problems constructed on the basis of the STABIL-data [56] and a reservoir-design problem given by A. Prékopa, see [30].

(5) <u>Reduced gradient method</u>.
> Computer used: IBM/PC AT.
> Phase-I: Maximizing the probability using the reduced gradient method.
> Computing F and ∇F: Monte-Carlo integration code of Szántai, see [65].
> Problems solved: STABIL [56] and a problem described in [66].
> Computing time: For the STABIL model with 52 constraints, 46 structural variables, 4-dimensional random right-hand-side and multinormal distribution the solution time was ~20 minutes on an IBM/PC AT without arithmetic coprocessor. This computing time corresponds to a solution which has been computed with an unnecessarily high accuracy lying much below 1%. For the test problem published in [66] the very simple Phase-I procedure implemented in the first version of the code turned out to be rather slow. The recent implementation called PROCON is an experimental version, the development of a robust implementation is in progress, along with a thorough computational study concerning efficiency and robustness.

Acknowledgement

I am indebted to one of the referees for a number of valuable comments and suggestions.

References

[1]   BAZARAA, M. S., SHETTY, C. M.: "Nonlinear programming. Theory and algorithms", *John Wiley & Sons* (1979).

[2]   BIRGE, J. R., WETS, R. J.-B.: "Designing approximation schemes for stochastic optimization problems, in particular for stochastic programs with recourse", *Math. Programming Study.* 27 (1986) 54-102.

[3]   BONNESEN, T., FENCHEL, W.: "Theorie der konvexen Körper", *Verlag von Julius Springer* (1934).

[4]   BORELL, C.: "Convex set-functions in d-space", *Periodica Math. Hungarica* 6 (1975) 111-136.

[5]   BOROS, E., PRÉKOPA, A.: "Closed form two-sided bounds for probabilities that at least r and exactly r out of n eventes occur", *Mathematics of Operations Res.* 14 (1989) 317-342.

[6]   BRASCAMP, H. J., LIEB, E. H.: "On extensions of the Brunn-Minkowski and Prékopa-Leindler theorems, including inequalities for log concave functions, and with an application to the diffusion equation", *J. of Functional Analysis* 22 (1976) 366-389.

[7]   BURKAUSKAS, A.: "On the convexity of probabilistic constrained stochastic programming problems", Thesis, *Hungarian Academy of Sciences*, Budapest (1984), (in Hungarian).

[8]   BURKAUSKAS, A.: "On the convexity of probabilistic constrained stochastic programming problems", *Alkalmazott Matematikai Lapok* 12 (1986) 77-90, (in Hungarian).

[9]   CHARNES, A., COOPER, W. W., SYMONDS, G. H.: "Cost horizons and certainty equivalents: An approach to stochastic programming of heating oil production", *Management Science* 4 (1958) 235-263.

[10]   CHARNES, A., COOPER, W. W.: "Chance constrained programming", *Management Science* 6 (1959) 73-89.

[11]   DEÁK, I.: "Three digit accurate multiple normal probabilities", *Numerische Mathematik* 35 (1980) 369-380.

[12]   DEÁK, I.: "Computing probabailities of rectangles in case of multinormal distributions", *J. Statist. Comp. and Simulation* 26 (1986) 101-114.

[13]   DEÁK, I.: "Multidimensional integration and stochastic programming", in Ermoliev, Y., Wets, R., J.-B., (eds.) *Numerical Techniques for Stochastic Optimization, Springer-Verlag*, Berlin (1988) 187-200.

[14]   DEMBO, R. S., KLINCEWICZ, J. G.: "Dealing with degeneracy in reduced gradient algorithms", *Math. Programming* 31 (1985) 357-363.

[15]   ERMOLIEV, Y.: "Stochastic quasigradient methods and their application to systems optimization", *Stochastics* 9 (1983) 1-36.

[16] FIACCO, A.V., McCORMICK, G.P.: "Nonlinear programming: Sequential unconstrained minimization technique", *John Wiley & Sons* (1968).

[17] GAIVORONSKI, A.: "Interactive program SQG-PC for solving stochastic programming problems on IBM/XT/AT compatibles - User Guide", *IIASA Working Paper* WP-88-11, (1988).

[18] GASSMANN, H.: "Conditional probability and conditional expectation of a random vector", in Ermoliev, Y., Wets, R., J.-B., (eds.) *Numerical Techniques for Stochastic Optimization, Springer-Verlag*, Berlin (1988) 237-254.

[19] GEOFFRION, A. M.: "Duality in nonlinear programming: A simplified applications-oriented approach", *SIAM Review* 13 (1971) 1-37.

[20] JOHNSON, N. L., KOTZ, S.: "Distributions in statistics: Continuous multivariate distributions", *Wiley* (1972).

[21] KALL, P.: "Qualitative Aussagen zu einigen Problemen der stochastischen Programmierung", *Z. Wahrscheinlichkeitstheorie und verw. Geb.* 6 (1966) 246-272.

[22] KALL, P. : "Stochastic linear programming", *Springer-Verlag*, Berlin, (1976).

[23] KALL, P.: "On approximation and stability in stochastic programming", in Guddat, J. et al. (eds.) *Parametric Optimization and Related Topics, Akademie-Verlag,* Berlin (1987) 387-407.

[24] KALL, P.: "A review on approximations in stochastic programming", *Preprint, IOR University of Zürich* (1989).

[25] KALL, P.: "Lösungsverfahren der stochastischen Programmierung - ein Ueberblick", in Kall, P., J. Kohlas, W. Popp, C.A. Zehner (eds.): *Quantitative Methoden in den Wirtschaftswissenschaften - Hans Paul Künzi zum 65. Geburtstag-, Springer- Verlag*, Berlin (1989) 19-29.

[26] KALL, P.: "Solution methods in stochastic programming - A review-", *Preprint, IOR University of Zurich* (1990).

[27] KALL, P., RUSZCZYNSKI, A., FRAUENDORFER, K.: "Approximation techniques in stochastic programming", in Ermoliev, Y., Wets, R., J.-B., (eds.) *Numerical Techniques for Stochastic Optimization, Springer-Verlag*, Berlin (1988) 33-64.

[28] KING, A., J.: "Stochastic programming problems: Examples from the literature", in Ermoliev, Y., Wets, R., J.-B., (eds.) *Numerical Techniques for Stochastic Optimization, Springer-Verlag*, Berlin (1988) 543-567.

[29] KLEINMICHEL, H., SADOWSKI, H.: "Der verallgemeinerte RG-Algorithmus bei linearen Restriktionen, die behandlung des Entartungsfalls und die Konvergenz des Verfahrens", *Beiträge zur Numerischen Mathematik* 3 (1975) 37-55.

[30] KOMÁROMI, É.: "A dual method for probabilistic constrained problems", *Math. Programming Study.* 28 (1986) 94-112.

[31] LUC, D. T.: "Duality in programming under probabilistic constraints with a random technology matrix", *Problems of Control and Information Theory* 12 (1983) 429-437.

[32] MADANSKY, A.: "Methods of solution of linear programs under uncertainty", *Operations Research* 10 (1962) 463-470.

[33] MARSTEN, R. E.: The design of the XMP linear programming library" *ACM Transactions on Mathematical Software* 7 (1981) 481-497.

[34] MARTI, K.: "Konvexitätsaussagen zum linearen Stochastischen Optimierungsproblem", *Z. Wahrscheinlichkeitstheorie und verw. Geb.* 18 (1971) 159-166.

[35] MAYER, J.: "A nonlinear programming method for the solution of a stochastic programming model of A. Prékopa", in Prékopa, A. (ed.) *Survey of Mathematical Programming, North-Holland,* Vol. 2 (1979) 129-139.

[36] MAYER, J.: "Probabilistic constrained programming: A reduced gradient algorithm implemented on PC", *IIASA Working Paper* WP-88-39 (1988).

[37] MILLER, B. L., WAGNER, H. M.: "Chance constrained programming with joint constraints", *Operations Research* 13 (1965) 930-945.

[38] MURTAGH, B. A., SAUNDERS, M. A.: "Large scale linearly constrained optimization", *Math. Programming* 14 (1978) 41-72.

[39] NEMHAUSER, G. L., WOLSEY, L. A.: "Integer and combinatorial optimization", *John Wiley & Sons,* (1988).

[40] NIEDERREITER, H.: "Quasi-Monte Carlo methods and pseudo-random numbers", *Bull. Amer. Math. Soc.* 84 (1978) 957-1041.

[41] van de PANNE, C., POPP, W.: "Minimum cost cattle feed under probabilistic problem constraints", *Management Sci.* 9 (1963) 405-430.

[42] PRÉKOPA, A.: "On probabailistic constrained programming", in *Proc. of the Princeton Symposium on Math. Programming, Princeton Univ. Press* (1970) 113-138.

[43] PRÉKOPA, A.: "Logarithmic concave measures with application to stochastic programming", *Acta. Sci. Math.* 32 (1971) 301-316.

[44] PRÉKOPA, A.: "A class of stochastic programming decision problems", *Math. Operationsforsch. Statist.,* 3 (1972) 349-354.

[45] PRÉKOPA, A.: "On logarithmic concave measures and functions", *Acta. Sci. Math.* 34 (1973) 335-343.

[46] PRÉKOPA, A.: "Contributions to stochastic programming", *Mathematical Programming* 4 (1973) 202-221.

[47] PRÉKOPA, A.: "Programming under probabilistic constraints with a random technology matrix", *Math. Operationsforsch. Statist., Ser. Optimization* 5 (1974) 109-116.

[48] PRÉKOPA, A.: "Eine Erweiterung der sogenannten Methode der zulässigen Richtungen der nichtlinearen Optimierung auf den Fall quasikonkaver Restriktionen", *Math. Operationsforsch. Statist., Ser. Optimization* 5 (1974) 281-293.

[49]   PRÉKOPA, A.: "Logarithmic concave measures and related topics", in Dempster. M. A. H., (ed.) *Stochastic Programming, Academic Press* (1980) 63-82.

[50]   PRÉKOPA, A.:"Numerical solution of probabilistic constrained programming problems", in Ermoliev, Y., Wets, R., J.-B., (eds.) *Numerical Techniques for Stochastic Optimization, Springer-Verlag*, Berlin (1988) 123-139.

[51]   PRÉKOPA, A.: "Boole-Bonferroni inequalities and linear programming", *Operations Research* 36 (1988) 145-162.

[52]   PRÉKOPA, A.: "Sharp bounds on probabilities using linear programming", *Operations Research* 38 (1990) 227-239.

[53]   PRÉKOPA, A.: "The discrete moment problem and linear programming", *Discrete Applied Math.* 27 (1990) 235-254.

[54]   PRÉKOPA, A.: "Dual method for the solution of a one-stage stochastic programming problem with random RHS obeying a discrete probability distribution", *ZOR,* to appear.

[55]   PRÉKOPA, A., BOROS, E. : "On the existence of a feasible flow in a stochastic transportation network", *Operations Research*, to appear.

[56]   PRÉKOPA, A., GANCZER, S., DEÁK, I., PATYI, K.: "The STABIL stochastic programming model and its experimental application to the electricity production in Hungary", in Dempster, M. A. H. (ed.): *Stochastic Programming, Academic Press*, London (1980) 369-385.

[57]   PRÉKOPA, A., KELLE, P.: "Reliability type inventory models based on stochastic programming", *Math. Programming Study* 9 (1983) 43-58.

[58]   PRÉKOPA, A., SZÁNTAI, T.: "Flood control reservoir system design using stochastic programming", *Math. Programming Study* 9 (1983) 138-151.

[59]   RAIKE, W. M.: "Dissection methods for solution in chance constrained programming problems", *Management Science* 16 (1970) 708-715.

[60]   RAPCSÁK, T.: "On the numerical solution of a reservoir model", *doctoral dissertation, University of Debrecen, Hungary* (1974) (in Hungarian).

[61]   RINOTT, Y.: "On convexity of measures", Annals of Probability 4 (1976) 1020-1026.

[62]   SALINETTI, G.: "Approximations for chance-constrained programming problems", *Stochastics* 10 (1983) 157-169.

[63]   SENGUPTA, J. K.: "Stochastic programming. Methods and applications", *North-Holland Publ. Co.* (1972).

[64]   SZÁNTAI, T.: "Evaluation of a special multivariate gamma distribution function", *Math. Programming Study* 27 (1986) 1-16.

[65]   SZÁNTAI, T.: "Calculation of the multivariate distribution function values and their gradient voctors", *IIASA Working Paper* WP-87-82 (1987).

[66]   SZÁNTAI, T.: "A computer code for solution of probabilistic-constrained stochastic programming problems", in Ermoliev, Y., Wets, R., J.-B., (eds.) *Numerical Techniques for Stochastic Optimization, Springer-Verlag*, Berlin (1988) 229-235.

[67]   ZOUTENDIJK, G.: "Nonlinear programming: A numerical survey", *J. SIAM Control* 4 (1966) 194-210.

STOCHASTIC OPTIMIZATION IN ACID RAIN
MANAGEMENT WITH VARIABLE METEOROLOGY

Edward A. McBean

Department of Civil Engineering
University of Waterloo
Waterloo, Ontario, Canada N2L 3G1

ABSTRACT

A multi-year linear programming model is developed to examine the impacts of variable meteorology on optimal controls for acid rain abatement in eastern North America. The features impacting the transfer coefficients relating emissions from sources to depositions at receptors, are discussed.

INTRODUCTION

Sulfur-based acid rain in eastern North America is caused primarily by anthropogenic emissions from the power and non-power industries. Due to the distributed nature of these emissions, the examination to find cost-effective emission reduction strategies for an area as large as eastern North American represents a complicated task in decision analyses. Complicating factors in finding a cost-effective emission reduction strategy include: numerous pollution sources and potential damage locations exist; and, there is a significant complexity governing the interrelationships between large-scale meteorologic phenomena and source-to-receptor transport of pollutants.

Any identified emission-reduction strategies for acid rain abatement must reflect several attributes: (i) they must be efficient in terms of minimizing control costs and (ii) they must be effective in terms of achieving desirable deposition objectives. In response, a number of different models have been developed (e.g. McBean et al (1985), Young and Shaw (1986), and Shaw (1986)).

The emission-reduction strategy models have played an important role in developing scenarios for decision-making in the real-world, by identifying cost-effective approaches (Donnan, 1989). However, much of the preliminary work was accomplished using deterministic models. Therefore, due to the stochastic nature of such phenomena as wind and precipitation fields, recent research work has been focussing on the development of stochastic components of the modelling (e.g. Ellis et al, 1986 and 1987; Ponnambalam et al 1990). The intent of this paper is to carry these works further, utilizing a multi-year formulation of the optimization model. Example results of the application of the model are included as a case study.

BACKGROUND

The identification of an effective and equitable means to mitigate acid rain problems is essential. An effective plan is necessary because the costs involved in reducing sulfur and sulfur oxides emissions are very large. For example, in Ellis et al (1985b), the potential savings attainable through optimized versus across-the-board abatement strategies were analyzed. The optimized approach, termed "emission-weighted removal in a political jurisdiction", essentially permits individual source removal levels to vary between zero and their upper bounds with the restriction that overall emission reductions in Canada and the U. S. equal 50 percent. The "emission-weighted" approach captures desirable source-receptor-specific, cost-effectiveness attributes (sources with relatively low marginal costs and relatively high contributions to areas of concern with deposition, exhibit high removal levels). Resulting costs, for comparable environmental quality in eastern North America, are summarized in Table 1.

Table 1 Summary Removal Costs for Several Strategies

Removal Costs x 10^9 $ (1981 US dollars)

	50 percent cut across-the-board in emissions	Emission-Weighted Removal = 50%
U.S.	62.1	21.9
Canada	4.3	1.9

Source: Ref. McBean (1987)

The results in Table 1 indicate that expenditures on a cost-effective basis are approximately 3 times as effective as administratively-simple, uniform (across-the-board) cutbacks.

However, the results indicated in Table 1 were generated using a deterministic linear programming (LP) model.

In mathematical terms, this deterministic LP model is written as

$$\text{Minimize costs as} \quad \sum_{j=1}^{N} C_j R_j \tag{1}$$

Subject to

$$\sum_{j=1}^{N} W_j (1-R_j) t_{ji} + \sum_{\ell}^{L} W_\ell t_{\ell i} + BG_i \le D \max, i \ \forall_i \tag{2}$$

where $C_j = C_j(R_j)$ a marginal cost of SO_2 removal at source j, $(j = 1, \ldots N)$ [$/kT ($SO_2$)];

R_j = SO_2 removal level at controllable source j (the R_j are the decision variables);

W_j = existing SO_2 emission rate [kT y^{-1}] at controllable source j;

t_{ji} = a unit transfer coefficient which represents the annual wet sulfur deposition rate [kg (Wet S) .ha^{-1}.y^{-1}], at receptor i $(i = 1, \ldots M)$, that results from one unit (1 kT (SO_2) y^{-1}) emission at source j;

W_ℓ = existing SO_2 emission rate (kT.y^{-1}) at noncontrollable source ℓ, $(\ell=1, \ldots L)$;

BG_i = background wet sulfur deposition rate [kg (Wet S).ha.$^{-1}$ y^{-1}];

$D_{max,i}$ = maximum allowable wet sulfur deposition rate [kg (Wet S) ha.$^{-1}$ y^{-1}].

The model formation implied in (1) and (2) can be easily augmented to include budgetary and equity constraints. In addition, because of the nonlinearity in the control costs, piecewise linear segments in the convex cost curves are typically employed wherein

$$R_j = \sum_{k=1}^{K} X_{jk} \tag{3}$$

for K piecewise linear segments at individual controllable sources j, $j = 1, \ldots N$.

Nevertheless, all of the principal elements of the deterministic model, namely the source emission rates, cost-removal functions, transfer coefficients, background deposition rates and deposition limits, can all be reasonably construed as random variables.

Interest herein will be restricted to the uncertainty in the transfer coefficients.

INCORPORATION OF UNCERTAINTY IN TRANSFER COEFFICIENTS

The principal difficulties associated with quantification of transfer coefficient uncertainty are related to the two-dimensional nature of long-range transport of acidic pollutants. Pollutant deposition at a particular location is the result of numerous source emissions, each a function of meteorologic conditions. Furthermore, these conditions exhibit considerable spatial and temporal variability.

One possible approach to capture the variabilities involves chance-constrained programming (CCP), in which the deterministic model is replaced by a more general probabilistic representation. Two such formulations were developed in Ellis et al (1985 (a) and 1986), namely:
(i) complete collinearity wherein strict dependence is assumed between all transfer coefficients; and,
(ii) complete noncollinearity in which strict independence is assumed between all transfer coefficients. These represent extreme upper and lower bounds of transfer coefficient uncertainty but the conservativeness of the assumptions/results diminishes their value.

As a more useful model, it is feasible to capture elements of the uncertainty through use of a multi-year model. Specifically, this can be accomplished as described below.

Defining $P_{\theta jT}$ as the probability of wind blowing from sector direction θ during year T, at location source j, then

$$\sum_{\theta=1}^{\theta} P_{\theta jT} = 1 \qquad (4)$$

In addition, the transfer linking source j to receptor i in year T for wind direction θ, $t_{jiT\theta}$, is potentially functionally dependent on precipitation at j and i and direction, in general terms as

$$t_{jiT\theta} = f \text{ (precip }_{jT}, \text{ precip}_{iT}, \text{ wind dispersion terms} = g\ (\theta)) \qquad (5)$$

where precip $_{kT}$ represents the precipitation at location k (for k = j and i) in year T and the wind dispersion terms quantifying the dispersive flux of the gaussian plume of pollutants, are potentially dependent on wind direction. However, Shipley et al (1987) showed that Equation (5) can be simplified to

$$t'_{jiT} \stackrel{\sim}{=} f \text{ (precip }_{jT}) \qquad (6)$$

In words, the transfer coefficient linking source j to receptor i in year T can be approximated as a function of the precipitation quantity at location j in year T.

Using (6), the transfer coefficient linking release at source j to deposition at receptor i during year T, can be written as

$$t_{jiT} = \sum_{\theta=1}^{\Theta} P_{\theta jT} t'_{ji} \ (\text{precip}_{jT}) \qquad (7)$$

The multi-year constraint equivalent to Equation (2) then becomes, for year T,

$$\sum_{j=1}^{N} [W_{jT} \ (1 - \sum_{k=1}^{K} X_{jkT}) \sum_{\theta=1}^{\Theta} P_{\theta jT} t'_{ji} (\text{precip}_{jT})] +$$

$$\sum_{\ell=1}^{L} [W_{\ell T} \sum_{\theta=1}^{\Theta} P_{\theta \ell T} t'_{\ell i} \ (\text{precip}_{jT})] + BG_i \leq D_{\max,i} \ \forall i, T \qquad (8)$$

The generality of (8) allows X_{jkT} to be a function of individual year T or, alternatively, with the imposition of

$$X_{jkT} = X_{jk} \quad \forall j, T \qquad (9)$$

the removal levels for all years T would be constrained to be identical.

CASE STUDY RESULTS

The sensitivity of the optimal controls with respect to variable meteorology were examined using the model described previously. Optimization was performed for each year independently using specification of $D_{\max,i} = 20$ kg/ha-yr. The transfer coefficients that relate emission from source j to deposition at receptor i in year t were estimated using the UW-LRT model (See Kompter and McBean, 1989) which has been shown to produce similar depositional patterns obtained by monitoring records over a series of years (1980-84).

The area being considered consists of all states of the U.S. bordering and to the east of the Mississippi River plus Texas and all the Canadian provinces east of British Columbia. The source inventories include 235 large emission sources identified on the basis that their 1980 emissions exceeded 19000 tonnes of SO_2 per year. These large sources represent approximately 67% of total man-made emissions of SO_2 that originated within the modelling area in 1980. The remaining emission sources (sources emitting less than 19000 tonnes per year and diffuse sources) were aggregated by state and province into the noncontrollable sources.

For each of the 235 large emission sources, SO_2 emission levels were specified along with estimated removal efficiencies and costs for alternative, new emission control techniques. Different meteorology (wind roses and precipitation levels) for different years (1980-84) was utilized to characterize the variability in abatement strategies in response to the varying meteorologic conditions.

The model was used to find the optional decision set independently for each of the years 1980-84 with varying transfer coefficients. With emission levels set at 1980 levels, the variability in solution is due strictly to the impact of

meteorologic variability. In Figure 1, the "optimal" cost is presented for each of
the transfer coefficient sets for 1980 through 1984 for a maximum deposition of
20 kg Ha-yr. For the maximum deposition limit of 20 kg wet-sulfate/ha-year, the
Canadian costs varied from 5.5 to 6.2 billion dollars, from the best year to the
worst, respectively, with a difference in costs of about 12%. The corresponding
cost variation for the United States is 37.4 to 51.3 billiion dollars, a difference
of about 38% between the best and worst years. Therefore, the cost variation is
much more significant for the United States than for Canada. The degree of
variability in costs for different years is, in part, a function of the stringency
of the deposition target; for low deposition targets (involving large removals and
costs), many of the available control options are implemented whereas for higher
allowable depositions, the needed/not needed implementation of several control
options will vary with different meteorological years. Thus, for a target of '20',
the high cost-effectiveness of the Canadian source controls, means that they are
always implemented before the implementation of the U.S. controls for achieving the
same deposition effect; thus the little variation in Canadian costs over the five
years.

For further examination of the specifics of cost effective strategies, the
interested reader is referred to, for example, Ponnambalam et al (1990).

CONCLUSIONS

LP models as a multi-year model can be useful tools to determine practical
solutions for the acid rain management problem. The uncertainties that exist in
transfer coefficients are extremely difficult to model explicitly because there are
variance and covariance terms across the variables and possibly across the
constraints as well. However, the multi-year optimization model captures the
variability, essential for examination of the impacts of meteorologic variability.

REFERENCES

Donnan, J., Policy and Planning Branch, Ontario Ministry of the Environment,
Toronto, Ontario, Personal Communications, 1989.

Ellis, J. H., McBean, E. A., and Farquhar, G. J., "Chance-Constraint/Stochastic
Linear Programming Model for Acid Rain Abatement - I. Complete Colinearity and
Noncolinearity", Atmospheric Environment, Vol. 19, No. 6, pp 925-937, 1985a.

Ellis, J. H., McBean, E. A., and Farquhar, G. J., "Chance-Constraint/Stochastic Linear Programm Model for Acid Rain Abatement - II. Limited Colinearity", Atmospheric Environment, Vol. 20, No. 3, pp 501-511, 1986.

Ellis, J. H., McBean, E. A., and Farquhar, G. J., "Deterministic Linear Programming Model for Acid Rain Abatement", ASCE-Journal of Environmental Engineering Division, III, pp 119-139, 1985 b.

Kompter, M. G., and McBean, E., "A Seasonal Long-Range Transport Model of Acid Pollutants with Preciptation Variability ", ASCE-Journal of Environmental Engineering, under review.

McBean, E. A., "Developing Management Positions for Acidic Emission Reduction Negotiations", in Systems Analysis in Water Quality Management, ed. M. Beck, Pergamon Press, 1987.

McBean, E., Ellis, J. H., and Fortin, M. "A Screening Model for Development and Evaluation of Acid Rain Abatement Strategies",Journal of Environmental Management, 21, pp 287-299, 1985.

Ponnambalam, K., McBean, E., and Unny, T., "Meteorological Impacts on Acid Rain Abatement Decisions", ASCE-Journal of Environmental Engineering Division, Vol. 116, No. 6, Nov./Dec. 1990.

Shaw, R., "A Proposed Strategy for Reducing Sulfate Deposition in North America, - II. Methodology for Minimizing Costs", Atmospheric Environment, 20 (1), pp 201-206, 1986.

Shipley, K., McBean, E., Farquhar, G., and Byrne, J., "Incorporation of Windroses in a Statistical Long-Range Pollution Transport Model", Water, Air and Soil Pollutions, 36, pp 115-130, 1987.

Young, J. W., and Shaw, R. W., "A Proposed Strategy for Reducing Sulfate Deposition in North America - I. Methodology for Minimizing Sulfur Removel", Atmospheric Environment, 20 (1), pp 189-199, 1986.

Figure 1. Variability in Control Costs for Different Meteorology, Given the Same Deposition Limitation

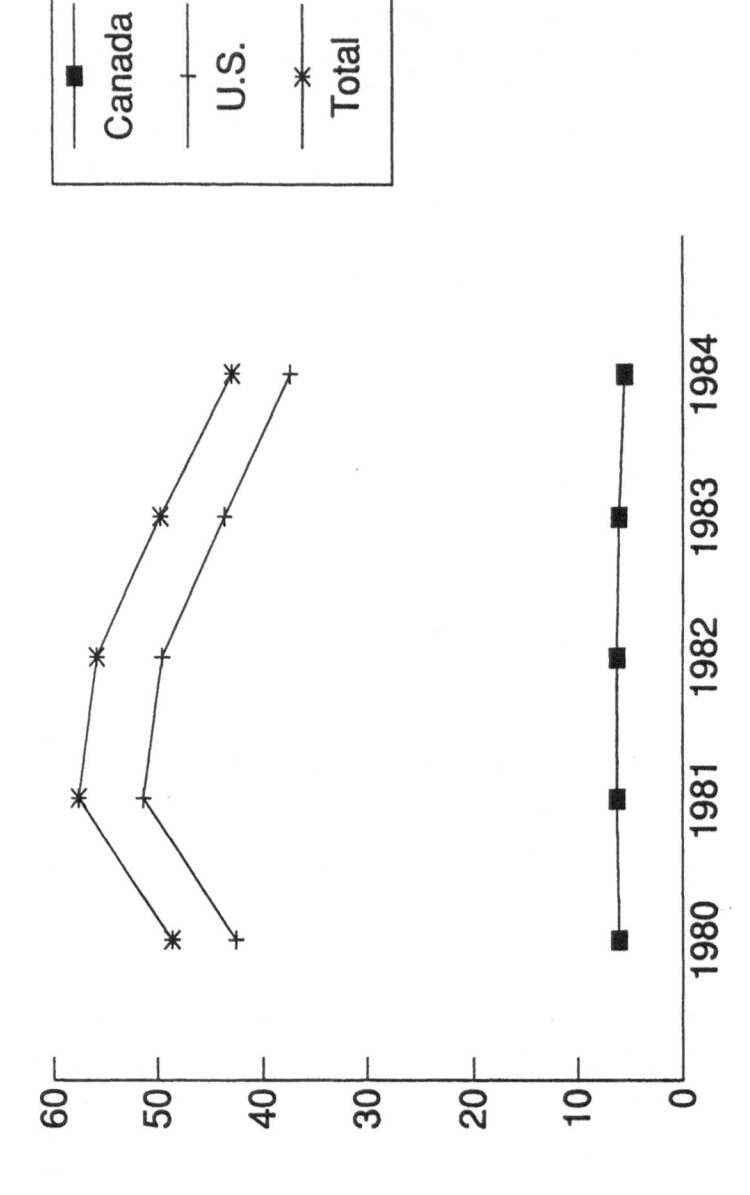

Deposition limit 20

COLLAPSE LOAD ANALYSIS AND OPTIMAL DESIGN BY STOCHASIC PROGRAMMING WITH UNCERTAINTIES OF LOADS

Anna Vásárhelyi

Technical University of Budapest Fac. Civil Engineering, Dep. Mathematics

H-1111 Budapest, Műegyetem rkp. 3.

Abstract

The effects of uncertainties of external loads are examined in case of plastic collapse load analysis and optimal design. Stochastic programming model can be written on base of kinematical theorems. In case of collapse load analysis Van de Panne and Popp's stochastic model of optimal design Kataoka's model are used. Applications are presented by analysis of large panel structures. The results are compared with the deterministic solutions.

I. Introduction

In structural analysis and design, two kinds of statistical uncertainties can occur; deviations of planned material quality [6] and of considered loads from actual values. In this paper, the effect of uncertainties of loads is examined, in the case of collapse load analysis and optimal plastic design. Taking into consideration the distribution function of loads the problems are modelled as stochastic programming problem. Application of the proposed method is illustrated by large panel structures.

In plastic analysis of structural elements, it is supposed that stresses are expressed at a finite number of points only and the yield condition of a structural element is represented by yield conditions at these points. In the following, rigid-ideal plastic material model is supposed because in limit state the elastic deformations can be neglected and the Tresca yield condition is used. The Tresca yield condition of an element is as follows:

$$\tau_i^l \leq \tau_i^m \leq \tau_i^u \quad i=1,\ldots,nm \quad , \tag{1}$$

where n is the number of points of an element,

m is the number of the elements,

τ_i^l, τ_i^u are the lower and upper yield limits in shear at i-th point of the element,

τ_i^m is the maximal main stress at i-th point of the element.

It is supposed the external loads have a multidimensional normal distribution function with given mean values and covariance matrix.

II. Plastic Collapse Load Analysis

The aim of plastic collapse load analysis is to determine the load carrying capacity of structures that are subjected to proportional loading. Considering only small strain, elastic deformations do not influence load carrying capacity. Therefore, it is sufficient to make analysis on basis of rigid-plastic structures. Assuming monotonically increasing loading, the structure arrives at the plastic limit state, characterized by the extreme value of the load parameter, only once.

Collapse load analysis follows two fundamental principles; the statical and the kinematical theorems which correspond to the primal-dual problems of linear programming, in deterministic case [1]. The statical theorem (the primal problem) in deterministic case can be formulated as follows:

$$G^* s + \alpha q = 0 \tag{2.a}$$
$$k^l \leq Ns \leq k^u \tag{2.b}$$
$$\alpha \to \max \tag{2.c}$$

where G^* is the transponant of the geometrical matrix of the structure,

N is the matrix of the linearized yield condition,

s is the vector of internal forces,

q is the load vector,

k^l, k^u are the lower and upper yield limit vectors, respectively,

α is the load parameter.

The load-carrying capacity of a structure corresponds to the maximum multiplier on applied loads (2.c), for which exists a statically admissible stress field (equilibrium equations are satisfied (2.a)), that ewerywhere satisfies the yield conditions (2.b).

According to the dual problem the kinematical theorem is formed:

$$G\ddot{u} + \lambda^{\cdot} N = 0 \tag{3.a}$$

$$q\dot{u} = 1 \tag{3.b}$$

$$\lambda^{\cdot}_{i} \geq 0 , \quad i=1,\ldots,nm \tag{3.c}$$

$$\alpha = k^{l}\lambda^{\cdot} + k^{u}\lambda^{\cdot} \to \min. \tag{3.d}$$

where $\dot{u}$ is the velocity vector

$\lambda^{\cdot}$ is the vector of plastic-rate multipliers.

The mechanical meaning of the dual problem is: the load-carrying capacity of a structure corresponds to the minimum multiplier α (3.d) on the applied loads for which there exists a kinematically admissible velocity field (the compatibility equations are satisfied (3.a)) that involves positive power of applied loads (3.b) and dissipation of mechanical energy (3.c).

To take into consideration the uncertainty of the external loads, start from the dual problem (3). The equ. (3.b) is demanded with a prescribed probability p. In this way it can be substituted with the following inequality:

$$P (q\dot{u} \leq 1) \geq p \tag{4}$$

If we consider for one component only:

$$P (q_{i}\dot{u}_{i} \leq 1) \leq 1 - p ,$$

introducing standardized variables:

$$P \left[\xi \leq \frac{1 - E(q_{i})\dot{u}_{i}}{\sqrt{\dot{u}_{i} C_{ii} \dot{u}_{i}}} \right] \leq 1 - p ,$$

$$\frac{1 - E(q_{i})\dot{u}_{i}}{\sqrt{\dot{u}_{i} C_{ii} \dot{u}_{i}}} \leq \Phi^{-1}(1 - p) ,$$

$$1 \leq E(q_{i})\dot{u}_{i} + \Phi^{-1}(1 - p) \sqrt{\dot{u}_{i} C_{ii} \dot{u}_{i}} ,$$

where E, C are the mean value and covariance, respectively,

Φ^{-1} is the inverse function of the standard normal distribution function.

In multidimensional case Van de Panne and Popp have proved [4] that a similar expression can be used if the number of dimensions is large.

$$1 \le \sum_{i=1}^{m} E(q_i)\dot{u}_i + \Phi^{-1}(1 - p) \sqrt{\dot{u}\, C\, \dot{u}} \tag{5}$$

Summarizing the mathematical programming model has to be solved is:

$$G\ddot{u} + \lambda^{\cdot} N = 0 \tag{6.a}$$

$$1 \le \sum_{i=1}^{m} E(q_i)\dot{u}_i + \Phi^{-1}(1 - p) \sqrt{\dot{u}\, C\, \dot{u}} = \beta \tag{6.b}$$

$$\lambda_i^{\cdot} \ge 0 \ , \ i=1,\ldots,n \tag{6.c}$$

$$k^l \lambda^{\cdot} + k^u \lambda^{\cdot} \to \min. \tag{6.d}$$

We do not save the equ. (3.b). Mechanically it means the external power is not normalized. It follows the value of $\alpha\beta$ is determined as the function value of the objective at the optimal point; that is the objective value has to be reduced by β for getting the load parameter.

III. Optimal Plastic Design

In cases of optimal plastic design, loads and geometry of the structure are given and values of yield limits at the discretized points in the structure are to be determined in such a way that weight of the structure be minimum. Usually it is supposed that weight of the structure is proportional to the sum of weighted values of yield limits. Optimal plastic design has two fundamental principles; the statical and kinematical theorems [1]. In deterministic case they corresponds to primal and dual problems of linear programming, as in case of collapse load analysis.

The statical theorem can be formulated as follows:

$$G^{*}s + \alpha q = 0 \tag{7.a}$$

$$k^l \le Ns \le k^u \tag{7.b}$$

$$\ell k \to \min. \tag{7.c}$$

where ℓ is the vectors of weighting values.

The dual problem express the kinematical theorem:

$$G\ddot{u} + \lambda N = 0 \tag{8.a}$$

$$0 \leq \lambda_i \leq \ell, \quad i=1,\ldots,n \tag{8.b}$$

$$q^* \dot{u} \to \max. \tag{8.c}$$

In stochastic case instead of the dual objective a new constrain and objective are introduced:

$$P \left(q^* \dot{u} \leq f \right) \geq p, \quad f \to \max.$$

where p is a prescribed probability level.

Using the former approximation (5) and eliminating f the stochastic programming model is:

$$G\ddot{u} + \lambda N = 0 \tag{9.a}$$

$$0 \leq \lambda_i \leq \ell, \quad i=1,\ldots,n \tag{9.b}$$

$$\sum_{i=1}^{m} E(q_i)\dot{u}_i + \Phi^{-1}(1 - p) \sqrt{\dot{u} \, C \, \dot{u}} \to \max. \tag{9.c}$$

It corresponds to the Kataoka's stochastic programming model [3] and he has proved it can be solved.

IV. Application to Large Panel Structures

Application of the methods described above are illustrated by analysis and design of prefabricated large panel structures. In solution, the rigid panel model proposed by Kaliszky is used [2]. According to this model, panels are considered as rigid bodies connected by springs that capable to transmitting tensile compressing and shear forces along panel edges (Fig.1). This means that between two adjacent panel edges, there are three springs; two to take tensile or compressing forces and one to take shear forces. It is supposed that the panels are constrained to remain in their plane.

Equilibrium equations for the (i,j)-th panel, which define the geometrical matrix, are as follows:

178

$$s_1^{(4)}(i,j) - s_1^{(2)}(i+1,j) + s_2^{(4)}(i,j) - s_2^{(2)}(i+1,j) + s_3^{(3)}(i,j) -$$

$$- s_3^{(3)}(i,j+1) + \alpha q_x(i,j) = 0$$

$$s_1^{(3)}(i,j) + s_2^{(3)}(i,j) - s_1^{(1)}(i,j+1) - s_2^{(1)}(i,j+1) + s_3^{(4)}(i,j) -$$

$$- s_3^{(2)}(i+1,j) + \alpha q_y(i,j) = 0$$

$$s_1^{(3)}(i,j)a + s_2^{(3)}(i,j)a + s_1^{(1)}(i,j+1)a - s_2^{(1)}(i,j+1)a - s_3^{(4)}(i,j)a +$$

$$+ s_3^{(2)}(i+1,j)a + s_1^{(4)}(i,j)b - s_2^{(4)}(i,j)b + s_1^{(2)}(i+1,j)b -$$

$$- s_2^{(2)}(i+1,j)b + s_3^{(3)}(i,j)b - s_3^{(1)}(i,j+1)b + \sigma M(i,j) = 0$$

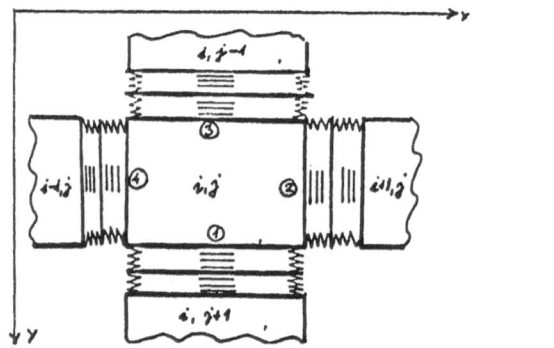 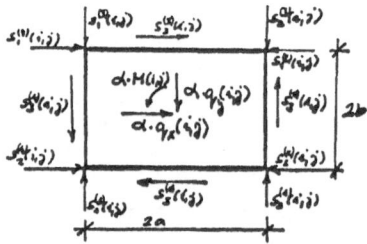

Figure 1.

Since the springs are in a uniaxial state of stress, the matrix of yield condition (N) is a diagonal matrix.

V. Examples

i. Collapse load analysis

It is examined how the covariance matrix influence the value of the load parameter. The example is given in Figure 2.

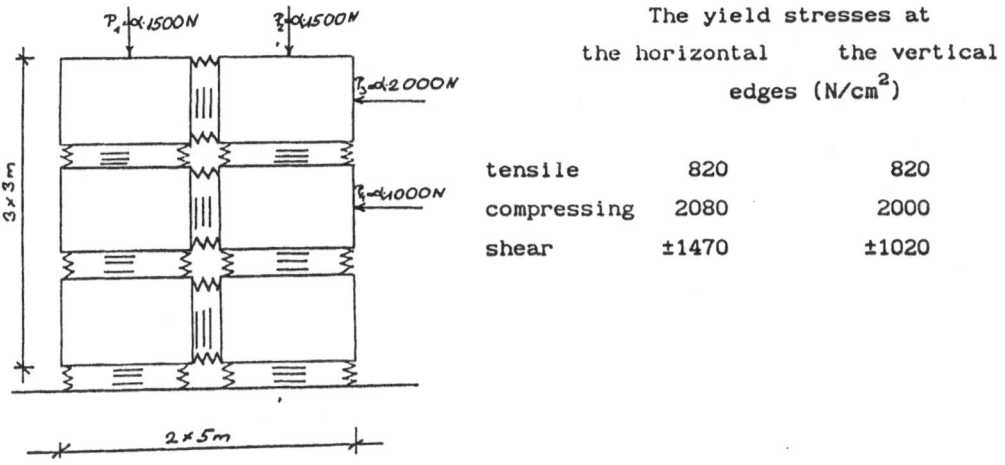

The yield stresses at

	the horizontal	the vertical
	edges (N/cm^2)	
tensile	820	820
compressing	2080	2000
shear	±1470	±1020

Figure 2.

The Figure 3. shows the solution in deterministic case, while the following examples concern to the stochastic solution with different covariances.

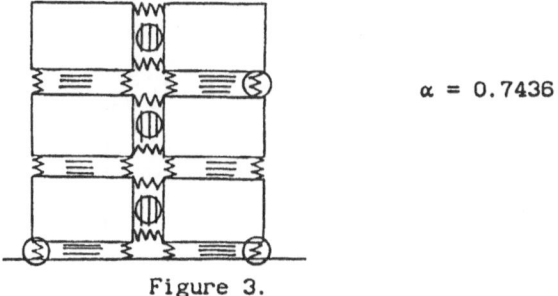

$\alpha = 0.7436$

Figure 3.

It can be seen that the average value of the load parameter is greater in stochastic cases than the deterministic solution, naturally, because the covariance matrix express the simultaneity of the loads. If the dependence among the loads are stronger, the load parameter is smaller.

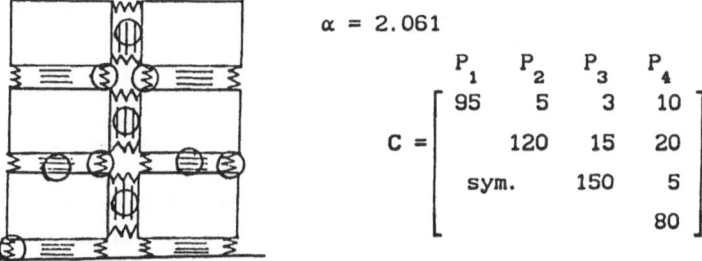

$\alpha = 2.061$

$$C = \begin{bmatrix} \overset{P_1}{95} & \overset{P_2}{5} & \overset{P_3}{3} & \overset{P_4}{10} \\ & 120 & 15 & 20 \\ \text{sym.} & & 150 & 5 \\ & & & 80 \end{bmatrix}$$

180

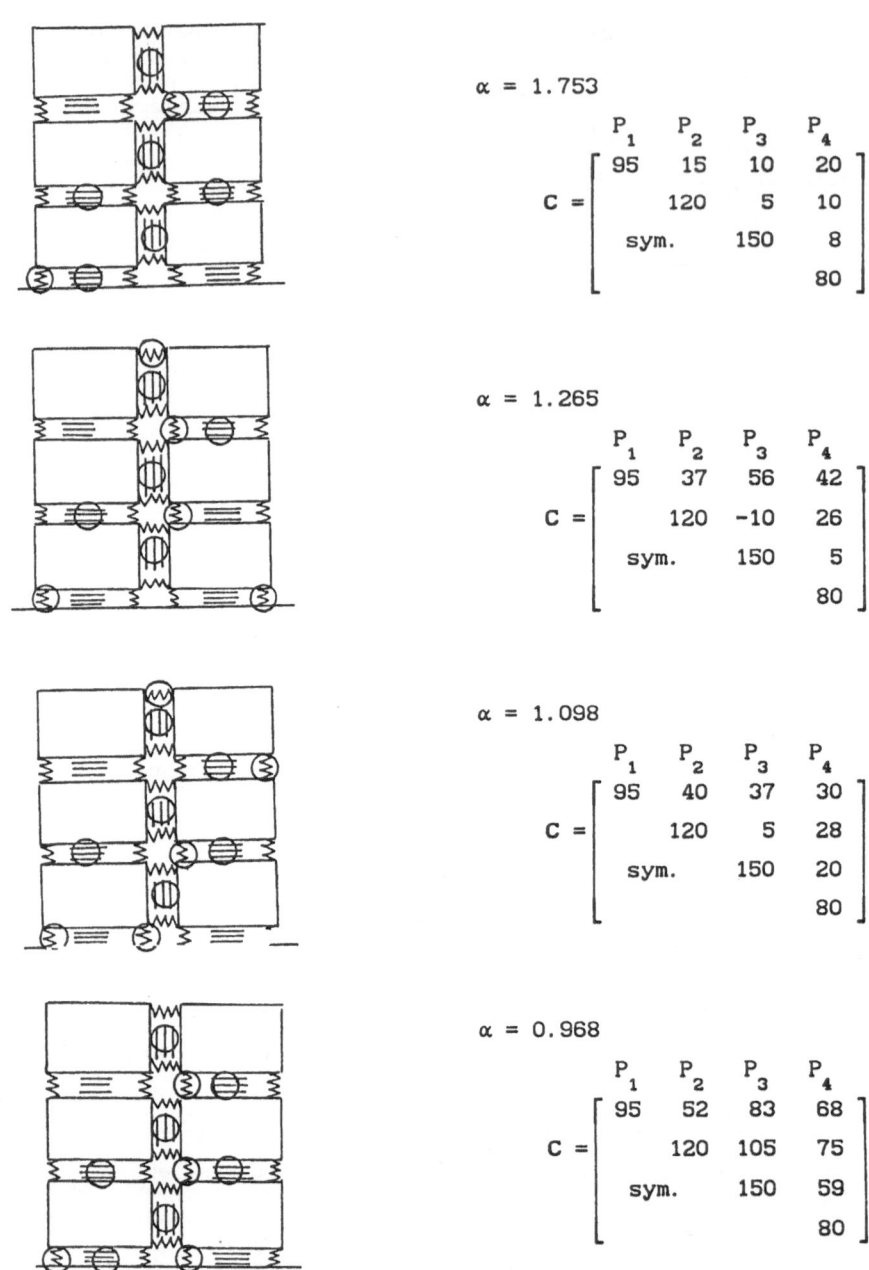

$$\alpha = 1.753$$

$$C = \begin{bmatrix} \overset{P_1}{95} & \overset{P_2}{15} & \overset{P_3}{10} & \overset{P_4}{20} \\ & 120 & 5 & 10 \\ \text{sym.} & & 150 & 8 \\ & & & 80 \end{bmatrix}$$

$$\alpha = 1.265$$

$$C = \begin{bmatrix} \overset{P_1}{95} & \overset{P_2}{37} & \overset{P_3}{56} & \overset{P_4}{42} \\ & 120 & -10 & 26 \\ \text{sym.} & & 150 & 5 \\ & & & 80 \end{bmatrix}$$

$$\alpha = 1.098$$

$$C = \begin{bmatrix} \overset{P_1}{95} & \overset{P_2}{40} & \overset{P_3}{37} & \overset{P_4}{30} \\ & 120 & 5 & 28 \\ \text{sym.} & & 150 & 20 \\ & & & 80 \end{bmatrix}$$

$$\alpha = 0.968$$

$$C = \begin{bmatrix} \overset{P_1}{95} & \overset{P_2}{52} & \overset{P_3}{83} & \overset{P_4}{68} \\ & 120 & 105 & 75 \\ \text{sym.} & & 150 & 59 \\ & & & 80 \end{bmatrix}$$

Figure 4.

The collapse mechanisms - denoted by circles on the Figure 3. and 4. - are very different, it is influenced by the covariances. In stochastic case the form of mechanisms can not be predicted.

181

ii. Optimal plastic design

Panels are considered as rigid bodies. Therefore, the weight of the structure depends on the connections only. For sake of simplicity, it was assumed that the objective function is the sum of the necessary yield stresses of the springs ($\ell=1$). The example is presented on Figure 5.

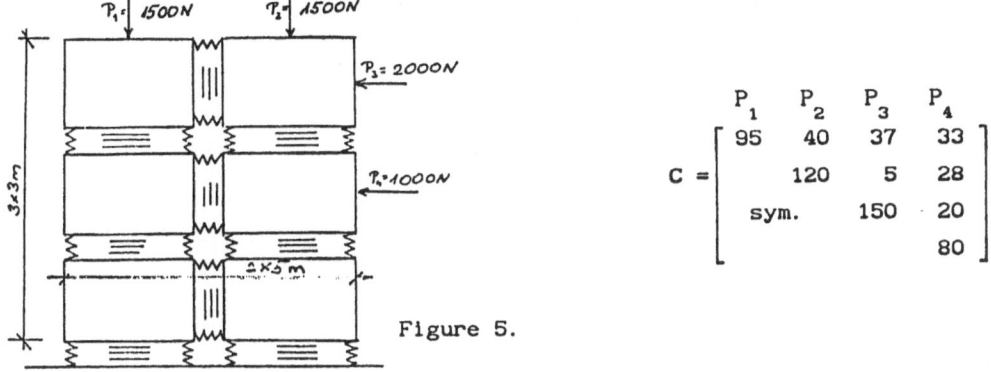

$$C = \begin{bmatrix} & P_1 & P_2 & P_3 & P_4 \\ & 95 & 40 & 37 & 33 \\ & & 120 & 5 & 28 \\ & \text{sym.} & & 150 & 20 \\ & & & & 80 \end{bmatrix}$$

Figure 5.

The Fig.6. and Fig.7. show the results in deterministic and stochastic case, respectively.

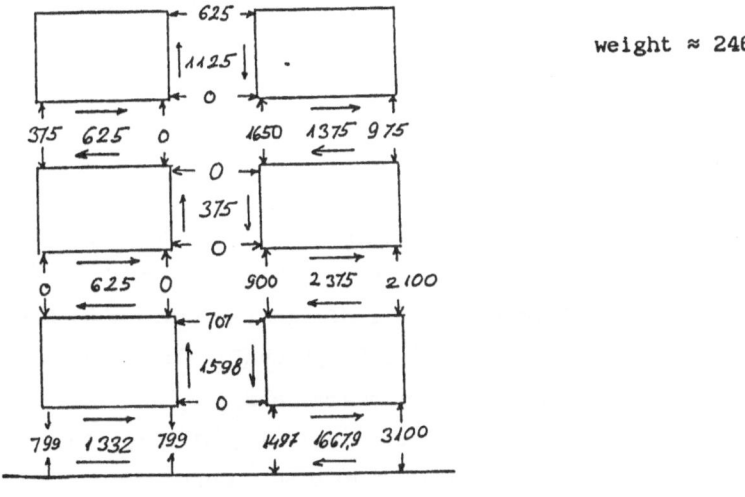

weight ≈ 2462.6

Figure 6.

The average value of the weight of the structure is a little less in stochastic case than in deterministic solution . The strongest spring is more feeble in stochastic than deterministic case and while in the

182

deterministic solution zero yield limits of the springs can be seen in the stochastic case the distribution of the yield limits is more uniform.

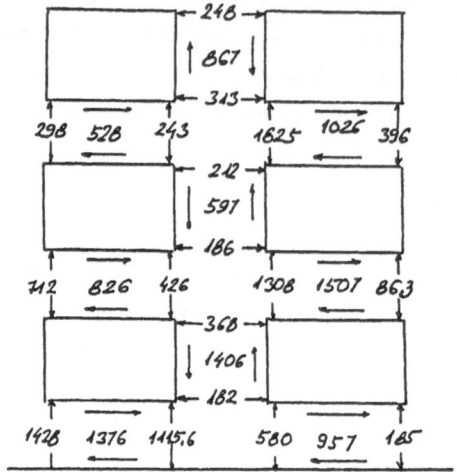

weight ≈ 1997.8

Figure 7.

To solve the problems above the Lagrange multiplier method was used [5]. The inverse of the one dimensional standard distribution function was approximated by its serie.

References

[1]. M. Z. Cohn and G. Maier, *Engineering Plasticity by Mathematical Programming*, Pergamon Press, New York, 1977.

[2]. S. Kaliszky and K. Wolf, *Analysis of Panel Buildings by Use of Rigid Models*, Periodica Polytechnica 23 (2):89-100, 1979.

[3]. I. Kataoka, *A Stochastic Programming Model*, Economica, Vol.31.(181-196), 1963

[4]. C. Van de Panne and W. Popp, *Minimum Cost Feed under Probabilistic Constraint*, Management Science, Vol 9, (43-51), 1963.

[5]. D. A. Pierre and M. A. Lowe, *Mathematical Programming via Augmented Lagrangians*, Addison-Wesley, London, 1975.

[6]. A. Vásárhelyi, *Limit Analysis and Optimal Plastic Design by Stochastic Programming with Uncertainties of Material's Quality*, Mech. Struc. & Mach. 15(2), (153-165), 1987.

Vol. 360: P. Stalder, Regime Translations, Spillovers and Buffer Stocks. VI, 193 pages . 1991.

Vol. 361: C. F. Daganzo, Logistics Systems Analysis. X, 321 pages. 1991.

Vol. 362: F. Gehreis, Essays In Macroeconomics of an Open Economy. VII, 183 pages. 1991.

Vol. 363: C. Puppe, Distorted Probabilities and Choice under Risk. VIII, 100 pages . 1991

Vol. 364: B. Horvath, Are Policy Variables Exogenous? XII, 162 pages. 1991.

Vol. 365: G. A Heuer, U. Leopold-Wildburger. Balanced Silverman Games on General Discrete Sets. V, 140 pages. 1991.

Vol. 366: J. Gruber (Ed.), Econometric Decision Models. Proceedings, 1989. VIII, 636 pages. 1991.

Vol. 367: M. Grauer, D. B. Pressmar (Eds.), Parallel Computing and Mathematical Optimization. Proceedings. V, 208 pages. 1991.

Vol. 368: M. Fedrizzi, J. Kacprzyk, M. Roubens (Eds.), Interactive Fuzzy Optimization. VII, 216 pages. 1991.

Vol. 369: R. Koblo, The Visible Hand. VIII, 131 pages.1991.

Vol. 370: M. J. Beckmann, M. N. Gopalan, R. Subramanian (Eds.), Stochastic Processes and their Applications. Proceedings, 1990. XLI, 292 pages. 1991.

Vol. 371: A. Schmutzler, Flexibility and Adjustment to Information in Sequential Decision Problems. VIII, 198 pages. 1991.

Vol. 372: J. Esteban, The Social Viability of Money. X, 202 pages. 1991.

Vol. 373: A. Billot, Economic Theory of Fuzzy Equilibria. XIII, 164 pages. 1992.

Vol. 374: G. Pflug, U. Dieter (Eds.), Simulation and Optimization. Proceedings, 1990. X, 162 pages. 1992.

Vol. 375: S.-J. Chen, Ch.-L. Hwang, Fuzzy Multiple Attribute Decision Making. XII, 536 pages. 1992.

Vol. 376: K.-H. Jöckel, G. Rothe, W. Sendler (Eds.), Bootstrapping and Related Techniques. Proceedings, 1990. VIII, 247 pages. 1992.

Vol. 377: A. Villar, Operator Theorems with Applications to Distributive Problems and Equilibrium Models. XVI, 160 pages. 1992.

Vol. 378: W. Krabs, J. Zowe (Eds.), Modern Methods of Optimization. Proceedings, 1990. VIII, 348 pages. 1992.

Vol. 379: K. Marti (Ed.), Stochastic Optimization. Proceedings, 1990. VII, 182 pages. 1992.